SEI/ASCE 24-98

Structural Engineering Institute
American Society of Civil Engineers

Flood Resistant Design and Construction

Published by the American Society of Civil Engineers
1801 Alexander Bell Drive
Reston, Virginia 20191-4400

ABSTRACT
This Standard, *Flood Resistant Design and Construction*, provides minimum requirements for flood-resistant design and construction of structures located in flood hazard areas. This standard applies to new construction which includes: (a) new structures including subsequent work to such structures, and (b) work classified as substantial repair or substantial improvement of an existing structure that is not an historic structure.

Library of Congress Cataloging-in-Publication Data

Flood resistant design and construction / Structural Engineering Institute, American Society of Civil Engineers.
    p. cm. — (ASCE standard)
"SEI/ASCE 24-98."
Includes bibliographical references and index.
ISBN 0-7844-0431-3
    1. Structural design. 2. Flood damage prevention—United States. 3. Building, Stormproof—United States.
I. Structural Engineering Institute. II. American Society of Civil Engineers.
TA658.2. F56   2000
721'.0467—dc21   00-022386

Photocopies. Authorization to photocopy material for internal or personal use under circumstances not falling within the fair use provisions of the Copyright Act is granted by ASCE to libraries and other users registered with the Copyright Clearance Center (CCC) Transactional Reporting Service, provided that the base fee of $8.00 per article plus $.50 per page is paid directly to CCC, 222 Rosewood Drive, Danvers, MA 01923. The identification for ASCE Books is 0-7844-0431-3/00/$8.00 + $.50 per page. Requests for special permission or bulk copying should be addressed to Permissions & Copyright Dept., ASCE.

Copyright © 2000 by the American Society of Civil Engineers,
All Rights Reserved.
Library of Congress Catalog Card No: 00-022386
ISBN 0-7844-0431-3
Manufactured in the United States of America.

STANDARDS

In April 1980, the Board of Direction approved ASCE Rules for Standards Committees to govern the writing and maintenance of standards developed by the Society. All such standards are developed by a consensus standards process managed by the Management Group F (MGF), Codes and Standards. The consensus process includes balloting by the balanced standards committee made up of Society members and nonmembers, balloting by the membership of ASCE as a whole, and balloting by the public. All standards are updated or reaffirmed by the same process at intervals not exceeding 5 years.

The following Standards have been issued.

ANSI/ASCE 1-82 N-725 Guideline for Design and Analysis of Nuclear Safety Related Earth Structures

ANSI/ASCE 2-91 Measurement of Oxygen Transfer in Clean Water

ANSI/ASCE 3-91 Standard for the Structural Design of Composite Slabs and ANSI/ASCE 9-91 Standard Practice for the Construction and Inspection of Composite Slabs

ASCE 4-98 Seismic Analysis of Safety-Related Nuclear Structures

Building Code Requirements for Masonry Structures (ACI 530-99/ASCE 5-99/TMS 402-99) and Specifications for Masonry Structures (ACI 530.1-99/ASCE 6-99/TMS 602-99)

ASCE 7-98 Minimum Design Loads for Buildings and Other Structures

ANSI/ASCE 8-90 Standard Specification for the Design of Cold-Formed Stainless Steel Structural Members

ANSI/ASCE 9-91 listed with ASCE 3-91

ASCE 10-97 Design of Latticed Steel Transmission Structures

SEI/ASCE 11-99 Guideline for Structural Condition Assessment of Existing Buildings

ANSI/ASCE 12-91 Guideline for the Design of Urban Subsurface Drainage

ASCE 13-93 Standard Guidelines for Installation of Urban Subsurface Drainage

ASCE 14-93 Standard Guidelines for Operation and Maintenance of Urban Subsurface Drainage

ASCE 15-98 Standard Practice for Direct Design of Buried Precast Concrete Pipe Using Standard Installations (SIDD)

ASCE 16-95 Standard for Load and Resistance Factor Design (LRFD) of Engineered Wood Construction

ASCE 17-96 Air-Supported Structures

ASCE 18-96 Standard Guidelines for In-Process Oxygen Transfer Testing

ASCE 19-96 Structural Applications of Steel Cables for Buildings

ASCE 20-96 Standard Guidelines for the Design and Installation of Pile Foundations

ASCE 21-96 Automated People Mover Standards—Part 1

ASCE 21-98 Automated People Mover Standards—Part 2

SEI/ASCE 23-97 Specification for Structural Steel Beams with Web Openings

ASCE 24-98 Flood Resistant Design and Construction

ASCE 25-97 Earthquake-Actuated Automatic Gas Shut-Off Devices

FOREWORD

The material presented in this Standard has been prepared in accordance with recognized engineering principles. This Standard should not be used without first securing competent advice with respect to their suitability for any given application. The publication of the material contained herein is not intended as a representation or warranty on the part of the American Society of Civil Engineers, or of any other person named herein, that this information is suitable for any general or particular use or promises freedom from infringement of any patent or patents. Anyone making use of this information assumes all liability from such use.

DEDICATION

The members of the Flood Resistant Design and Construction Standards Committee of the Structural Engineering Institute respectfully dedicate this Standard in the memory of Raymond R. Fox, P.E., F.ASCE. His engineering expertise and professional dedication to the task of preparing this new standard guided the efforts of all committee members. His wise counsel and leadership will be greatly missed.

ACKNOWLEDGMENTS

The American Society of Civil Engineers (ASCE) acknowledges the devoted efforts of the Flood Resistant Design and Construction Standards Committee of the Structural Engineering Institute. This group comprises individuals from many backgrounds including: consulting engineering, research, construction industry, education, government, design, and private practice.

Work on this new standard was initiated in December, 1995 and performed through the consensus standards process in compliance with the procedures of ASCE's Rules for Standards Committees. Those individuals who serve on the Flood Resistant Design and Construction Standards Committee of SEI are:

Paul Armstrong
Conrad Battreal
Nageshwar R. Bhaskar
Brian A. Casement
Michael A. Cassaro
William L. Coulbourne
Anne M. Ellis
Shou Shan Fan
Kenneth A. Ford
Raymond R. Fox
John W. Gaythwaite
James S. Graham, Jr.
David Greenwood
Vernon K. Hagen
Philip H. Hasselwander
Barbara D. Hayes
Cheryl A. Johnson
Christopher P. Jones
Young C. Kim

Edward M. Laatsch
Thomas MacAllen
Samuel L. McNair
Joseph J. Messersmith, Jr.
Jimmy S. O'Brien
Clifford Oliver
Kim Paarlberg
Spencer M. Rodgers, Jr.
Herbert S. Saffir
Edward M. Salsbury
Phillip J. Samblanet
Erez Sela
Harry B. Thomas, Chair
Theodore C. Van Kirk
Richard A. Vognild
Robert A. Wessel
Wallace A. Wilson

Task Committee "A"
General Provisions, Flood Hazard Areas, Documents

Nageshwar R. Bhaskar
Vernon K. Hagen
Cheryl A. Johnson
Christopher P. Jones, Chair
Clifford Oliver

Task Committee "B"
High Risk Flood Hazard Areas, Flood Hazard Areas Subject to High Velocity Wave Action

David Greenwood, Chair
Philip H. Hasselwander
Thomas MacAllen
Spencer M. Rogers, Jr.

Task Committee "C"
Design, Materials, Accessory Structures

William L. Coulbourne
Raymond R. Fox, Chair
Joseph J. Messersmith, Jr.

Task Committee "D"
Dry and Wet Flood Proofing, Utilities, Egress

Conrad Battreal
Clifford Oliver, Chair
Theodore C. Van Kirk

CONTENTS

	Page
STANDARDS	iii
FOREWORD	v
DEDICATION	vii
ACKNOWLEDGMENTS	ix

- 1.0 General 1
  - 1.1 Scope 1
  - 1.2 Definitions 1
  - 1.3 Identification of Flood Hazard Areas 6
  - 1.4 Loads in Flood Hazard Areas 6
    - 1.4.1 General 6
    - 1.4.2 Combination of Loads 6
  - 1.5 Identification of Floodprone Structures 6
    - 1.5.1 General 6
    - 1.5.2 Consideration for Flood Protective Works 6
  - 1.6 Classification of Structures 7

- 2.0 Basic Requirements for Flood Hazard Areas 7
  - 2.1 Scope 7
  - 2.2 General 7
  - 2.3 Siting 8
    - 2.3.1 Siting in Flood Hazard Areas 8
    - 2.3.2 Siting in Floodways 8
    - 2.3.3 Siting in High Risk Flood Hazard Areas 8
  - 2.4 Elevation Requirements 8
    - 2.4.1 Flood Hazard Areas Not Subject to High Velocity Wave Action 8
    - 2.4.2 Flood Hazard Areas Subject to High Velocity Wave Action 8
  - 2.5 Foundation Requirements 8
    - 2.5.1 Geotechnical Considerations 8
    - 2.5.2 Foundation Depth 9
    - 2.5.3 Use of Fill 9
      - 2.5.3.1 Structural Fill 9
      - 2.5.3.2 Non-Structural Fill 9
    - 2.5.4 Use of Load Bearing Walls 9
      - 2.5.4.1 Required Openings in Load Bearing Foundation Walls 9
      - 2.5.4.2 Openings in Breakaway Walls 9
    - 2.5.5 Use of Piers, Posts, Columns, or Piles 9
  - 2.6 Enclosures Below the Design Flood Elevation 10
    - 2.6.1 Openings in Enclosures Below the Design Flood Elevation 10
      - 2.6.1.1 Non-Engineered Openings in Enclosures Below the Design Flood Elevation 10
      - 2.6.1.2 Engineered Openings in Enclosures Below the Design Flood Elevation 10

- 3.0 High Risk Flood Hazard Areas 11
  - 3.1 Scope 11
  - 3.2 Alluvial Fan Areas 11
    - 3.2.1 Protective Works in Alluvial Fan Areas 11
  - 3.3 Flash Flood Areas 11
    - 3.3.1 Protective Works in Flash Flood Areas 11

	3.4	Mudslide Areas	11
		3.4.1 Protective Works in Mudslide Areas	12
	3.5	Erosion Prone Areas	12
		3.5.1 Protective Works in Erosion Prone Areas	12
	3.6	High Velocity Flow Areas	12
		3.6.1 Protective Works in High Velocity Flow Areas	12
	3.7	High Velocity Wave Action Areas	12
	3.8	Icejam and Debris Areas	12
		3.8.1 Protective Works in Icejam and Debris Areas	12
4.0	Flood Hazard Areas Subject to High Velocity Wave Action		12
	4.1	Scope	12
		4.1.1 Identification of Areas Subject to High Velocity Wave Action	13
	4.2	General	13
	4.3	Siting	13
	4.4	Elevation Requirements	13
	4.5	Foundation Requirements	13
		4.5.1 General	13
		4.5.2 Special Geotechnical Considerations	15
		4.5.3 Foundation Depth	15
		4.5.4 Use of Fill	15
		4.5.5 Pile Foundations	15
		4.5.5.1 Piles Terminating in Caps at or Below Grade	15
		4.5.5.2 Piles Extending to Superstructure (Structure Framing)	16
		4.5.5.3 Wood Piles	16
		4.5.5.4 Steel H Piles	16
		4.5.5.5 Concrete-Filled Steel Pipe Piles and Shells	16
		4.5.5.6 Prestressed Concrete Piles	16
		4.5.6 Columns	16
		4.5.6.1 Wood Posts	16
		4.5.6.2 Reinforced Masonry Columns	16
		4.5.6.3 Reinforced Concrete Columns	16
		4.5.7 Grade Beams	16
		4.5.8 Bracing	17
	4.6	Enclosed Areas Below Design Flood Elevation	17
		4.6.1 Breakaway Walls	17
	4.7	Erosion Control Structures	17
	4.8	Decks, Concrete Pads, and Patios	17
5.0	Design		18
	5.1	General	18
	5.2	Vertical Structural Systems	18
		5.2.1 Masonry Walls	18
		5.2.2 Concrete Walls	18
		5.2.3 Lateral Resistance of Open Foundation Systems	18
		5.2.4 Piles	18
		5.2.4.1 General	18
		5.2.4.2 Timber Piles	18
		5.2.4.3 Steel HP-Section Piles	18
		5.2.4.4 Concrete-Filled Steel Pipe Piles and Shells	18
		5.2.4.5 Precast (Including Prestressed) Concrete Piles	18
		5.2.4.6 Cast-in-Place Concrete Piles	18

		5.2.4.7	Pile Capacity	18
		5.2.4.8	Capacity of the Supporting Soils	19
		5.2.4.9	Minimum Penetration	19
		5.2.4.10	Foundation Pile Spacing	19
		5.2.4.11	Pile Caps	19
		5.2.4.12	Timber Pile Connections	19
		5.2.4.13	Timber Piles Not in Tension	19
		5.2.4.14	Timber Piles in Tension	19
		5.2.4.15	Steel Piles Not in Tension	19
		5.2.4.16	Steel Piles in Tension	19
		5.2.4.17	Concrete Piles Not in Tension	20
		5.2.4.18	Concrete Piles in Tension	20
		5.2.4.19	Pile Splicing	20
		5.2.4.20	Mixed Types of Piling and Multiple Types of Installation Methodology	20
	5.2.5	Posts, Piers, and Columns		20
		5.2.5.1	Application	20
		5.2.5.2	Reinforced Concrete and Masonry Columns	20
5.3	Footings			20
5.4	Mats, Rafts, and Slabs			20
5.5	Grade Beams			20
5.6	Anchorage/Connections			20

6.0 Materials 21
    6.1   General 21
    6.2   Specific Materials Requirements for Flood Hazard Areas 21
        6.2.1   Metal Connectors and Fasteners 21
        6.2.2   Structural Steel 21
            6.2.2.1   Corrosive Environments 21
            6.2.2.2   Non-Corrosive Environments 21
        6.2.3   Concrete 22
        6.2.4   Masonry 22
        6.2.5   Wood and Timber 22
        6.2.6   Finishes 22

7.0 Dry and Wet Floodproofing 22
    7.1   Scope 22
    7.2   Dry Floodproofing 22
        7.2.1   Dry Floodproofing Restrictions 22
        7.2.2   Dry Floodproofing Requirements 23
    7.3   Wet Floodproofing 23
        7.3.1   Wet Floodproofing Restrictions 23
    7.4   Active Floodproofing 23

8.0 Utilities 23
    8.1   General 23
    8.2   Electrical 24
        8.2.1   Service Conduits and Cables 24
        8.2.2   Exposed Conduits and Cables 24
        8.2.3   Electric Meters 24
        8.2.4   Disconnect Switches and Circuit Breakers 24
        8.2.5   Electric Elements Installed Below Minimum Elevations 24

	8.3	Plumbing	25	
		8.3.1 Buried Plumbing Systems	25	
		8.3.2 Exposed Plumbing Systems	25	
		8.3.3 Plumbing Systems Installed Below Minimum Elevations	25	
		8.3.4 Sanitary Systems	25	
	8.4	Mechanical, Heating, Ventilation, and Air Conditioning	25	
		8.4.1 Fuel Storage Tanks	25	
	8.5	Elevators	26	
9.0	Means of Egress		26	
	9.1	General	26	
	9.2	Stairs and Ramps	26	
10.0	Accessory Structures		26	
	10.1	General	26	
	10.2	Decks, Porches, and Patios	26	
	10.3	Garages	27	
		10.3.1 Garages Attached to Structures	27	
		10.3.2 Detached Garages	27	
	10.4	Chimneys and Fireplaces	27	
11.0	References		27	

Commentary

C1.0	General	29
C2.0	Basic Requirements for Flood Hazard Areas	33
C3.0	High Risk Flood Hazard Areas	37
C4.0	Flood Hazard Areas Subject to High Velocity Wave Action	40
C5.0	Design	46
C6.0	Materials	47
C7.0	Dry and Wet Floodproofing	48
C8.0	Utilities	49
C9.0	Means of Egress	50
C10.0	Accessory Structures	50
C11.0	References	51
Index		55

Flood Resistant Design and Construction

1.0 GENERAL

1.1 SCOPE

This Standard provides minimum requirements for flood-resistant design and construction of structures located in flood hazard areas. This Standard applies to new construction which includes: (a) new structures including subsequent work to such structures, and (b) work classified as substantial repair or substantial improvement of an existing structure that is not an historic structure (see Fig. 1-1).

1.2 DEFINITIONS

The following definitions apply to the provisions of the entire Standard: (*Italicized* words in a definition mean the words are defined in this section.)

A Zone—An area within the *Special Flood Hazard Area*, which is not subject to high velocity wave action.

Adjacent Grade—The elevation of the natural or regraded ground surface, or *structural fill*, at the location of a structure.

Alluvial Fan—Fan-shaped deposits of sediment eroded from steep slopes and watersheds and deposited on valley floors.

Alluvial Fan Flooding—A type of flood hazard that occurs only on *alluvial fans*. Alluvial fan flooding is considered hazardous when designated as a *flood hazard area* on a communities flood hazard map or otherwise legally designated.

Apex—The highest point on an *alluvial fan* or similar landform, where the flow is last confined. The apex generally corresponds to the location where the *watershed* erosion ceases and fan sediment deposition commences.

Attendant Utilities and Equipment—Utilities, plumbing, HVAC and related equipment and services associated with new construction.

Authority Having Jurisdiction—The organization, community, political subdivision, office or agency which has adopted this Standard under due legislative authority.

Base Flood Elevation—The elevation in relation to the datum specified on a community's FIRM expected to be reached by a flood having a 1% chance of being equaled or exceeded in any given year.

Base Flood—The flood having a 1% chance of being equaled or exceeded in any given year.

Basement—That portion of a *structure* having its lowest floor below ground level on all sides.

Bedrock—Rock, usually solid, that underlies soil or other unconsolidated surficial material.

Braided Channel—A stream that is characterized by relatively shallow interlaced channels divided by islands or bars.

Breakaway Wall—Any type of wall using materials and construction techniques approved by the *authority having jurisdiction*, which does not provide structural support to a *structure*, and which is designed and constructed to fail under specified circumstances without damage to the structure, or supporting foundation system.

Bulkhead—A wall or structure to retain or prevent sliding or erosion of the land; sometimes used to protect against wave action.

Channel—A natural or artificial waterway which periodically or continuously contains moving water.

Channel Stabilization—Use of bank protection or other measures to prevent a channel from changing its size and shape, or from migrating laterally.

Check Valve—A valve that permits water or fluids to pass in one direction, but automatically closes when water or fluid attempts to flow through the valve in the opposite direction, also known as a backflow preventer.

Coastal High Hazard Area (CHHA)—An area within an area of special flood hazard extending from offshore to the inland limit of a primary frontal dune along an open coast and any other area which is subject to high velocity wave action from storms or seismic sources. This area is designated on FIRM's as velocity zones V, VO, VE, or V1-30.

Community—Any State or area or political subdivision thereof, or any Indian tribe or authorized tribal organization, or Alaska Native village or authorized native organization, which has the authority to adopt and enforce this Standard for areas within its jurisdiction.

Conveyance—The capacity of a channel to conduct flow, also referred to as the carrying capacity of the channel.

Debris Flow—A mass movement of sediment, including boulders, organic materials and other debris; debris flows typically move in surges, and are characterized by a steep frontal wave.

Debris Impact Loads—Loads on a *structure* caused by floodborne debris striking the structure, or

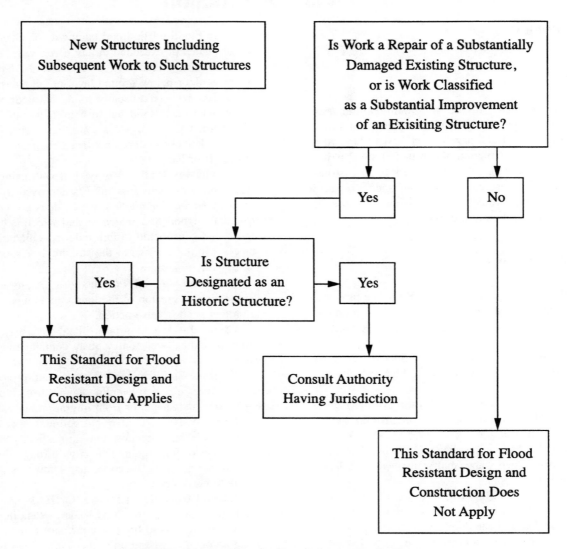

FIGURE 1-1. Diagram for Application of this Standard.

a portion thereof, often sudden in nature and large in magnitude.

Design Flood—The *flood* associated with the greater of the following two areas: 1) area within a *floodplain* subject to a 1% or greater chance of flooding in any year, or 2) area designated as a *flood hazard area* on a community's flood hazard map, or otherwise legally designated.

Design Flood Elevation (DFE)—The elevation of the *design flood*, including *wave height,* relative to the National Geodetic Vertical Datum (NGVD), North American Vertical Datum (NAVD), or other specified datum.

Dry Floodproofing—A floodproofing method used to render a *structure* envelope *substantially impermeable* to the entrance of floodwater.

Enclosure—A confined area below the *DFE*, formed by walls on all sides of the enclosed space.

Eroded Ground Elevation—The elevation of ground following any *erosion* expected during the *design flood* event.

Erodible Soil—Soil subject to wearing away and movement due to the effects of wind, water or other geological processes, during a flood or storm or over a period of years.

Erosion—The wearing away of the land surface by detachment and movement of soil and rock fragments, during a flood or storm or over a period of years, through the action of wind, water or other geologic processes.

Erosion Analysis—An analysis of the *erosion* potential of soil or strata, to include the effects of flooding or storm surge, moving water, wave action and the interaction of water and structural components.

Existing Structure—Any *structure* for which the *start of construction* commenced before the effective date of the first floodplain management code, ordinance or standard adopted by the *authority having jurisdiction*.

Fetch—The distance over which wind acts on the water surface to generate waves.

Fill—Material such as soil, gravel or crushed stone which is placed in an area to increase ground elevations (see *Structural Fill*).

Flap Gate—A gate placed at the end of a pipe, culvert or conduit, which permits water to be discharged but which prevents water from entering the discharge end of the pipe, culvert or conduit.

Flash Flood—A *flood* that crests in a short length of time and is often characterized by *high velocity flow*; it often results from heavy rainfall over a localized area, which overflows a confined water course. A flood whose waters rise from within banks flow to 3 ft or more above banks in less than 2 h shall be considered a flash flood.

Flood or Flooding—A general and temporary condition of partial or complete inundation of normally dry land from: 1) the overflow of inland or tidal waters, or 2) the unusual and/or rapid accumulation of runoff or surface waters from any source.

Flood Control Structure—A barrier designed and constructed to keep water away from or out of a specified area (see *Flood Protective Works*).

Flood-Damage-Resistant Material—Any construction material capable of withstanding direct and *prolonged contact with floodwaters*, without sustaining any damage that requires more than cosmetic repair.

Flood Fringe—That portion of the floodplain outside the floodway that serves as a temporary storage area for flood waters during a flood; flood waters in the flood fringe are normally shallower and of lower velocity than those of the floodway. The flood fringe is also known as the floodway fringe.

Flood Hazard Area—The greater of the following two areas: 1) *Special Flood Hazard Area*; or 2) area designated as a flood hazard area on a community's flood hazard map, or otherwise legally designated.

Flood Hazard Map—The map delineating *flood hazard areas* adopted by the authority having jurisdiction.

Flood Hazard Study—The study that serves as the technical basis for a *flood hazard map*.

Flood Protective Works—Barriers designed and constructed to keep water away from or out of a specified area (see *Flood Control Structure; Floodwall*).

Flood-Related Erosion—Collapse, subsidence or wearing away of land as a result of the action of flooding, including the effects of storm surge, moving water and wave action.

Flood Routing—The process of tracking the flood hydrograph, by calculation, including the course, magnitude and duration of a flood as it progresses through a channel, floodplain system (including *alluvial fans*) or reservoir area.

Floodplain—Any land area, including watercourse, susceptible to partial or complete inundation by water from any source.

Floodproofing—Any combination of structural or nonstructural adjustments, changes or actions which reduce or eliminate flood damage to a *structure*, contents, *attendant utilities and equipment*.

Floodwall—A constructed barrier of *flood-damage-resistant materials* used to keep water away from or out of a specified area.

Floodway—That portion of the *floodplain*, designated by the *authority having jurisdiction*, set aside to convey the greatest portion of the flow during a flood; obstructions in the floodway will act to reduce conveyance and increase flood elevations.

Footing—The enlarged base of a foundation, wall, pier or column, designed to spread the load of the *structure* so that it does not exceed the soil bearing capacity.

Footprint—The horizontal extent of a *structure*.

Freeboard—An additional height used as a factor of safety in determining the elevation of a *structure*, or *floodproofing*, to compensate for factors that may increase the flood height.

General Scour—During flood conditions, the removal of material from all or a major portion of the channel cross-section or land surface.

High Risk Flood Hazard Area—A *flood hazard area* where one or more of the following hazards are known to occur: *alluvial fan flooding, flash floods, mudslides, icejams, high velocity flows, high velocity wave action*, and *erosion*.

High Velocity Flow—During *design flood* or lesser conditions, water movement adjacent to *structures* and/or foundations with flow velocities greater than 10 ft/s.

High Velocity Wave Action—Condition present

in the *Coastal High Hazard Area*, where wave heights are greater than or equal to 3.0 ft in height, or where wave runup elevations reach 3.0 ft or more above grade.

Highest Adjacent Grade—The highest elevation of the natural or regraded ground surface, or *structural fill*, at the location of a *structure*.

Historic Structure—Any *structure* that meets one of the following criteria: 1) listed individually in the National Register of Historic Structures, 2) certified by the Secretary of the Interior as meeting the requirements for individual listing in the National Register, 3) certified or preliminary determination by the Secretary of the Interior as contributing to the historical significance of a registered historic district or a district preliminarily determined by the Secretary to qualify as a registered historic district, 4) individually listed on a state inventory of historic places, in states with historic preservation programs which have been approved by the Secretary of the Interior, or 5) individually listed on a local inventory of historic places in communities with historic programs certified by an approved state program or by the Secretary of the Interior.

Human Intervention—The required presence and active involvement of people to enact a flood-proofing measure prior to flooding.

Hydrodynamic Loads—Loads imposed on an object by water flowing against and around it.

Hydrostatic Loads—Loads imposed on an object by a standing mass of water.

Icejam—An accumulation of floating ice fragments that causes the bridging or damming of a channel or stream.

Impact Loads—Loads which result from debris, ice and any object transported by floodwaters striking against *structures* or parts thereof.

Levee—A man-made barrier, usually an earthen embankment, designed and constructed in accordance with sound engineering practices, to contain, control or divert the flow of water so as to provide protection from temporary flooding.

Local Scour—During flood conditions, the removal of material from a localized portion of the channel cross-section or land surface, due to an abrupt change in flow direction or velocity around an object or structural element.

Lowest Floor—The lowest floor of the lowest enclosed area, including basements; however, an unfinished or flood-resistant enclosure used solely for parking, building access, or storage shall not be considered the lowest floor.

Mangrove Stand—An assemblage of mangrove trees containing one or more of the following species: black mangrove, red mangroves, white mangrove or buttonwood.

Mud Flood—Hyperconcentrated sediment flow with a sediment concentration less than 45% by volume. Characterized by distinct fluid properties in deformation, particle setting, wave motion and spreading on a horizontal surface. A mud flood is a turbulent flood phenomenon.

Mudflow—Hyperconcentrated sediment flow with a sediment concentration in excess of 45% by volume, characterized by plastic deformation, cohesive characteristics, and lack of wave motion and spreading on a horizontal surface. A mudflow is a viscous flow.

Mudslide—General category of hyperconcentrated sediment flows including *mudflows*, *mud floods*, and *debris flows*.

NAVD—North American Vertical Datum.

New Construction—New structures including subsequent work to such structures, and work classified as substantial repair or substantial improvements of an existing structure that is not a historical structure.

New Structure—Any structure for which *start of construction* commenced after the effective date of the first floodplain management code, ordinance, or standard adopted by the *authority having jurisdiction*.

NGVD—National Geodetic Vertical Datum.

Non-Erodible Soil—Soil not subject to wearing away or movement due to the effects of wind, water or other agents during a flood or storm or a period of years.

Obstruction—Any object or structural component not permitted by this standard, attached to a *structure* below the DFE, that can cause an increase in flood elevation, deflect flood waters or transfer flood loads to any structure.

One-Hundred Year Flood—The *flood* having a 1% chance of being equaled or exceeded in any given year; also known as the 1% flood or 1% chance flood.

Pile—A structural element that is embedded in soils by drilling, driving or jetting, so that axial loads in the member are supported through skin friction or end bearing with the soil or rock and lateral loads in the member are supported through side bearing with the soil or rock.

Prolonged Contact with Floodwaters—Partial or total inundation by floodwaters for more than 72 h.

Rapid Drawdown—Rapid lowering of flood elevation, at a rate equal to or exceeding 5 ft/h.

Rapid Rise—Rapid increase in flood elevation, at a rate equal to or exceeding 5 ft/h.

Regulatory Floodway—Any *floodway* referenced in a floodplain management ordinance for the purpose of applying regulations.

Relocation—Moving a *structure* from a *flood hazard area* to a location of reduced flood hazard or to a location outside the *floodplain*.

Sand Dune—Natural or artificial ridges or mounds of sand landward of a beach.

Scour—The removal of soil or fill material from the channel cross-section or land surface by the flow of flood waters.

Seawall—A wall separating land and water areas, primarily designed to prevent erosion and other damage due to wave action.

Shear Wall—A load bearing or non-load bearing wall that transfers by in-plane lateral forces, lateral loads acting on a *structure*, to its foundation.

Shield—A removable or permanent watertight protective cover for an opening in a *structure* below the *DFE*, used in *floodproofing* the *structure*.

Special Flood Hazard Area—Land in the floodplain subject to a 1% or greater chance of flooding in any given year; area delineated on the Flood Insurance Rate Map as Zone A, AE, A1-30, A99, AR, AO, AH, V, VO, VE, or V1-30.

Start of Construction—Means the date the construction permit was issued for *new construction*, provided that actual start of construction, commenced within 180 days of the permit date. The actual start means either the first placement of permanent construction of a *structure* on a site, such as the pouring of a slab or footing, the installation of piles, the construction of columns, or any other work beyond the stage of excavation; or the placement of a manufactured home. Permanent construction does not include land preparation, such as clearing, grading or filling; nor does it include excavation for a basement, footings, piers, or foundation or the erection of temporary forms; nor does it include the installation of accessory structures, such as garages or sheds not occupied as dwelling units or not part of the main structure. For *substantial repair* or *substantial improvement*, the actual start of construction means the first alteration of any wall, ceiling, floor, or other structural part of a structure, whether or not that alteration affects the external dimensions of the structure.

Stillwater Depth—The vertical distance between the ground and the *stillwater elevation*.

Stillwater Elevation—The elevation that the surface of the water would assume in the absence of waves referenced to *NAVD*, *NGVD*, or other datum.

Storage Tank—A closed vessel used to store gases or liquids.

Structural Fill—*Fill* compacted to a specified density to provide structural support or protection to a *structure*.

Structure—Any building or other *structure* to which requirements of this standard are deemed applicable by the *authority having jurisdiction*, and gas and liquid storage tanks that are principally above ground.

Substantial Damage—Damage of any origin sustained by a *structure*, whereby the cost of restoration to its pre-damage condition equals or exceeds 50% of its pre-damage market value, or equals or exceeds a smaller percentage established by the *authority having jurisdiction*.

Substantial Improvement—Any reconstruction, rehabilitation, addition or other improvement to a *structure*, the cost of which equals or exceeds 50% of its pre-improvement market value, or equals or exceeds a smaller percentage established by the *authority having jurisdiction*.

Substantial Repair—Repair of a *substantially damaged* structure to which the requirements of this standard apply.

Substantially Impermeable—Use of *flood-damage-resistant materials* and techniques for *dry floodproofing* portions of a *structure*, which result in a space free of through cracks, openings, or other channels that permit unobstructed passage of water and seepage during flooding, and which result in a maximum accumulation of 4 in. of water depth in such space during a period of 24 h.

V Zone—Velocity Zones V, VO, VE, or V1-30 (see *Coastal High Hazard Area*).

Water Surface Elevation—The elevation of the water surface, usually referenced to *NGVD*, *NAVD* or another datum.

Watershed—A topographically defined area drained by a river or stream, or by a system of connecting rivers and streams such that all outflow is discharged through a single outlet.

Wave—A ridge, deformation or undulation of the water surface.

Wave Crest Elevation—The elevation of the crest of a *wave*, usually referenced to *NGVD*, *NAVD* or another datum.

Wave Height—The vertical distance between the crest and the trough of a *wave*.

Wave Loads—Loads imparted on a *structure* caused by waves striking the structure, or a portion thereof.

Wave Runup—The rush of wave water running up a slope or *structure*.

Wave Runup Elevation—The elevation, usually referenced to *NGVD, NAVD* or another datum, reached by *wave runup*.

Wet Floodproofing—A *floodproofing* method which relies on the use of *flood-damage-resistant materials* and construction techniques to minimize flood damages to areas below the *DFE*, of a *structure* intentionally allowed to flood (see *Floodproofing*).

1.3 IDENTIFICATION OF FLOOD HAZARD AREAS

This Standard shall apply to the larger of the two areas listed below:

1. Those lands within a floodplain subject to a 1% or greater chance of flooding in any year (i.e., the area subject to flooding during the base flood event, if such an event has been defined for a community).
2. Those lands designated as a flood hazard area on a community's flood hazard map, or otherwise legally designated if a community regulates to a higher standard than the 1% flood or base flood.

The flood associated with the governing definition listed above shall be termed the design flood. Design and construction requirements for new structures shall be dictated by conditions during the design flood.

1.4 LOADS IN FLOOD HAZARD AREAS

1.4.1 General

Design of structures within flood hazard areas shall be governed by the loading provisions of ASCE-7 *Minimum Design Loads for Buildings and Other Structures* [1].

Design and construction of structures located in flood hazard areas shall consider all flood-related hazards, including: hydrostatic loads, hydrodynamic loads, wave action; debris impacts; alluvial fan flooding; flood-related erosion; iceflows or icejams; or mudslides in accordance with requirements of this standard if specified or if not specified in this standard then in accordance with requirements approved by the authority having jurisdiction. Design documents shall identify and take into account flood-related and other concurrent loads that will act on the structure. Design documents shall include, but not be limited to, the applicable conditions listed below:

1. wave action
2. high velocity flood waters
3. impacts due to debris in the flood waters
4. rapid inundation by flood waters
5. rapid drawdown of flood waters
6. prolonged inundation by flood waters
7. wave and flood-induced erosion and scour
8. deposition and sedimentation by flood waters

1.4.2 Combination of Loads

Flood loads shall be combined with other concurrent loads as specified in the provisions for combinations of loads of ASCE-7 [1] either by using the allowable stress design method load combinations or by using the strength design method load combinations.

1.5 IDENTIFICATION OF FLOODPRONE STRUCTURES

1.5.1 General

A determination shall be made as to whether or not a structure lies, in whole or in part, within a flood hazard area following review of flood hazard maps, studies available in public domain and other information available from the authority having jurisdiction.

1.5.2 Consideration for Flood Protective Works

Dams, levees, floodwalls, diversions, channels and other flood protective works shall not be considered to provide protection for structures during the design flood, unless those works have been determined to provide protection during design flood conditions by the authority having jurisdiction. Design of structures behind levees and floodwalls determined to provide protection during the design flood shall consider trapping of rainfall, runoff and other waters behind the works, and shall include provision for drainage of interior areas during the design flood.

New construction behind flood protective works determined to provide protection during design flood conditions shall not:

1. damage, endanger or otherwise harm the protective works, or be in conflict with maintenance and

**TABLE 1-1. Classification of Structures for Flood Resistant Design and Construction
(Classification same as ASCE 7 [1])**

Nature of Occupancy	Category
Structures that represent a low hazard to human life in the event of failure including, but not limited to: • agricultural facilities[a] • certain temporary facilities • minor storage facilities	I
All structures except those listed in Categories I, III and IV	II
Structures that represent a substantial hazard to human life in the event of failure including, but not limited to: • Structures where more than 300 people congregate in one area • Structures with elementary school, secondary school, or day-care facilities with capacity greater than 250 • Structures with a capacity greater than 500 for colleges or adult education facilities • Healthcare facilities with a capacity of 50 or more resident patients but not having surgery or emergency treatment facilities • Jails and detention facilities • Power generating stations and other public utility facilities not included in Category IV • Structures containing sufficient quantities of toxic or explosive substances to be dangerous to public if released	III
Structures designated as essential facilities including but not limited to: • Hospitals and other health-care facilities having surgery or emergency treatment facilities • Fire, rescue, and police stations and emergency vehicle garages • Designated earthquake, hurricane, or other emergency shelters • Communications centers and other facilities required for emergency response • Power generating stations and other public utility facilities required in an emergency • Structures having critical national defense functions	IV

[a]Certain agricultural structures may be exempt from some of the provisions of this Standard—see Section C1.6

repair operations of the authority having jurisdiction
2. significantly increase the potential for trapping of rainfall, runoff and other waters behind the protective works

1.6 CLASSIFICATION OF STRUCTURES

Structures shall be classified according to Table 1-1.

2.0 BASIC REQUIREMENTS FOR FLOOD HAZARD AREAS

2.1 SCOPE

The requirements of Section 2 shall apply to new construction in flood hazard areas, unless more restrictive requirements are specified in Section 3 for new construction in High Risk Flood Hazard Areas and/or in Section 4 for new construction in Flood Hazard Areas subject to High Velocity Wave Action.

2.2 GENERAL

New construction shall be designed, constructed, connected and anchored to resist flotation, collapse or permanent lateral movement resulting from the action of hydrostatic, hydrodynamic, wind and other loads during design flood, or lesser, conditions in accordance with requirements of this Standard if specified, or if not specified in this Standard then in accordance with requirements approved by the authority having jurisdiction. Design shall include those loads and load combinations described in Sections 1.4.1 and 1.4.2.

Design and construction in flood hazard areas shall account for each of the following:

1. elevation of the structure relative to the design flood elevation (DFE)
2. foundations and geotechnical factors
3. obstructions or enclosures below the DFE
4. structural connections

FLOOD RESISTANT DESIGN AND CONSTRUCTION

5. use of flood damage resistant materials
6. floodproofing
7. utilities
8. means of egress

2.3 SITING

2.3.1 Siting in Flood Hazard Areas
New construction in flood hazard areas shall be designed, constructed and situated to minimize damage to the structure up to and during the design flood, and to minimize adverse impacts to other structures and property.

2.3.2 Siting in Floodways
Structures shall not be sited in floodway areas without demonstrating that those structures will not, during the design flood: 1) increase the flood level above that permitted by the authority having jurisdiction, and 2) reduce the conveyance of the floodway below that permitted by the authority having jurisdiction.

2.3.3 Siting in High Risk Flood Hazard Areas
Structures shall not be sited in high risk flood hazard areas, unless the additional requirements contained in Sections 3 and 4 of this Standard are met.

2.4 ELEVATION REQUIREMENTS

2.4.1 Flood Hazard Areas Not Subject to High Velocity Wave Action
Structures shall have the lowest floor (including basements) elevated to or above the DFE in conformance with the requirements of Table 2-1, excluding enclosed areas that are used solely for parking, building access or storage unless the lowest floor below the minimum elevation specified by Table 2-1 meets the floodproofing requirements of Section 7 of this Standard.

2.4.2 Flood Hazard Areas Subject to High Velocity Wave Action
Structures shall have the bottom of the lowest horizontal structural member of the lowest floor elevated to or above the DFE, in conformance with the requirements of Table 4-1, which has higher minimum elevation requirements than Table 2-1. Other than breakaway walls, no exceptions to the elevation requirements of Table 4-1 shall be made in areas subject to high velocity wave action.

TABLE 2-1. Minimum Elevation of Lowest Floor Relative to Base Flood Elevation (BFE) or Design Flood Elevation (DFE)—Flood Hazard Areas Not Subject to High Velocity Wave Action[a]

Structure Category[b]	Minimum Elevation of Lowest Floor[c]
I	at DFE
II	at DFE
III	at DFE
IV	at BFE + 1 ft or DFE whichever is higher

[a] Minimum elevations shown in Table 2-1 do not apply to areas subject to high velocity wave action (see Table 4-1). Minimum elevations shown in Table 2-1 do not apply to other high risk flood hazard areas, where specific elevation requirements are given in Section 3 of this Standard.
[b] See Table 1-1 for structure descriptions.
[c] Lowest floor shall be permitted below the minimum elevation if the structure meets the floodproofing requirements of Section 7.

2.5 FOUNDATION REQUIREMENTS

Foundations of structures shall be designed and constructed to support the structures during design flood conditions, and shall provide the required support to prevent flotation, collapse or permanent lateral movement under the load combinations specified in Section 1.4.2. Any part of the structure that provides structural support to that portion above the elevation specified by Table 2-1 or 4-1 shall meet all foundation requirements in this Standard.

2.5.1 Geotechnical Considerations
Foundation design shall be based on the geotechnical characteristics of the soils and strata below the structure, and on interactions between the soils and strata and the foundation. Foundation design shall account for instability and decreased structural capacity associated with: soil consolidation, expansion or movement; erosion and scour; liquefaction; and subsidence.

Geotechnical information necessary to complete the foundation design shall be obtained through geotechnical investigations of the site or from existing available data, such as investigations conducted at nearby project sites, regional studies conducted by government agencies or other reliable sources.

2.5.2 Foundation Depth

The foundation shall extend to a depth based on geotechnical considerations to provide the support described in Sections 2.5 and 2.5.1 above, taking into account the erosion and local scour of the supporting soil based on an erosion analysis.

2.5.3 Use of Fill

Structural fill, and non-structural fill, shall not be placed in floodway areas if such placement causes an increase in the flood level during the design flood event.

2.5.3.1 Structural Fill

Structural fill shall not be used in areas subject to high velocity wave action or in other high risk flood hazard areas unless design and construction of the structural fill account for:

1. consolidation of the underlying soil under the weight of the fill and the structure
2. differential settlement due to variations in fill composition and characteristics
3. slope stability and erosion control

Foundations in areas subject to high velocity wave action or in other high risk flood hazard areas must conform with the requirements of Section 3 and 4 of this Standard.

Fill used for structural support or protection shall be suitable for its intended use. Fill used to support or protect a structure shall be placed in lifts of not more than 12-in. loose thickness, with each lift compacted to at least 95% of its maximum Standard Proctor density (see [2]), unless a soils engineering report approved by the authority having jurisdicition specifies otherwise.

The side slopes of structural fill shall be no steeper than 1 on 1.5 (vertical/horizontal). Structural fill, including side slopes, shall be protected from scouring and erosion under flood conditions up to and including the design flood.

2.5.3.2 Non-Structural Fill

Use of non-structural fill beneath a structure shall be prohibited, except for minimal site grading to meet community drainage requirements.

2.5.4 Use of Load Bearing Walls

Load bearing walls extending below the DFE shall be designed and constructed to account for:

1. hydrostatic, hydrodynamic, soil, wind and other lateral loads acting during design flood conditions
2. buoyancy, dead load, live load and other vertical loads acting during design flood conditions
3. strength, stability and capacity of footings, other supporting structural elements and underlying soils
4. strength, stability and properties of materials used for bearing wall construction
5. connections to footings and other supporting structural elements

Load bearing walls extending below the DFE in areas subject to high velocity wave action or in other high risk flood hazard areas shall also meet the requirements of Sections 3 and 4.

2.5.4.1 Required Openings in Load Bearing Foundation Walls

Load bearing foundation walls that enclose an area below the DFE, and that do not meet the dry-floodproofing requirements of Section 7.2, shall contain openings to allow for automatic entry and exit of floodwaters during design flood conditions. These openings shall meet the requirements of Section 2.6.1.

2.5.4.2 Openings in Breakaway Walls

Openings to allow for the automatic entry and exit of floodwaters during design flood conditions shall be installed in breakaway walls in A Zones. The openings shall meet the requirements of Section 2.6.1. Openings in breakaway walls in V Zones shall not be required.

2.5.5 Use of Piers, Posts, Columns, or Piles

Piers, posts, columns or piles used to elevate a structure above the DFE in flood hazard areas shall comply with all applicable foundation requirements of this Standard. Connections between mat or raft foundations and piers, posts and columns shall meet all applicable foundation requirements of this Standard.

Design of pier, post, column or pile foundations shall account for:

1. hydrostatic, hydrodynamic, soil, wind and other lateral loads acting during design flood and lesser flood conditions
2. buoyancy, dead load, live load and other vertical loads acting during design flood and lesser flood conditions
3. strength, stability and capacity of underlying soils, and footings or other supporting structural elements, if present

FLOOD RESISTANT DESIGN AND CONSTRUCTION

4. strength, stability and properties of materials used for foundation construction
5. connections to footings or other supporting structural elements, if present

2.6 ENCLOSURES BELOW THE DESIGN FLOOD ELEVATION

Non-load bearing walls and other enclosures in flood hazard areas not subject to high velocity wave action or in other high risk flood hazards shall meet the dry-floodproofing requirements of Section 7.2, or contain openings to allow for automatic entry and exit of floodwaters during design flood and lesser flood conditions. Openings shall meet the requirements of Section 2.6.1.

Non-load bearing walls and other enclosures in flood hazard areas subject to high velocity wave action and in other high risk flood hazard areas shall meet the breakaway requirements of Section 4.6.1. Openings for the entry and exit of floodwaters shall not be required in breakaway walls located in flood hazard areas subject to high velocity wave action.

2.6.1 Openings in Enclosures Below the Design Flood Elevation

2.6.1.1 Non-Engineered Openings in Enclosures Below the Design Flood Elevation

Non-engineered openings shall meet the following criteria:

1. there shall be a minimum of two openings on different sides of each enclosed areas; if a structure has more than one enclosed area below the DFE, each area shall have openings on exterior walls
2. the total net area of all openings shall be at least 1 in.^{2} for each square foot of enclosed area
3. the bottom of each opening shall be no more than 1 ft above the adjacent ground level
4. openings shall not be less than 3.0 in. in diameter
5. any louvers, screens or other opening covers shall not block or impede the automatic flow of floodwaters into and out of the enclosed areas
6. openings meeting requirements 1 through 5 above installed in doors and windows are acceptable; however, doors and windows are not deemed to meet the requirements of this Standard.

2.6.1.2 Engineered Openings in Enclosures Below the Design Flood Elevation

Engineered openings shall meet the following criteria:

1. Each individual opening, and any louvers, screens or other covers, shall be designed to allow automatic entry and exit of floodwaters during design flood or lesser flood conditions.
2. There shall be a minimum of two openings on different sides of each enclosed area; if a structure has more than one enclosed area below the DFE, each area shall have openings.
3. Openings shall not be less than 3.0 in. in diameter.
4. The bottom of each required opening shall be no more than 1 ft above the adjacent ground level.
5. The difference between the exterior and interior floodwater levels shall not exceed 1 ft during periods of maximum rate of rise and maximum rate of fall of the floodwaters, and at other times during the design, or lesser, flood events.
6. In the absence of reliable data on the rates of rise and fall, assume a minimum rate of rise and fall of 5.0 ft/h; where an analysis indicates the rates of rise and fall are greater than 5.0 ft/h, the total net area of the required openings shall be increased to account for the higher rates of rise and fall; where an analysis indicates the rates of rise and fall are less than 5.0 ft/h, the total net area of the required openings shall remain the same or shall be decreased to account for the lower rates of rise and fall.
7. The minimum total net area of the required openings in non-breakaway walls shall be calculated using the equation:

$$A_o = 0.033(1/c)(R)(A_e)$$

where

A_o = total net area of openings required (in.^{2});
0.033 = coefficient (in.^{2}/h/ft^3) corresponding to a factor of safety of 5.0;
R = worst case rate of rise and fall (ft/h);
A_e = total enclosed area (ft^2); and
c = opening coefficient of discharge given in Table 2-2.

8. The minimum total net area of the required openings in breakaway walls shall be calculated using the equation:

$$A_o = 0.007(1/c)(R)(A_e)$$

where

A_o = total net area of openings required (in.^{2});

TABLE 2-2. Flood Opening Coefficient of Discharge

Opening Shape and Condition	c
Circular, unobstructed during design flood	0.60
Square, unobstructed during design flood	0.35
Rectangular, long axis horizontal, short axis vertical, unobstructed during design flood	0.40
Rectangular, short axis horizontal, long axis vertical, unobstructed during design flood	0.25
Other shapes, unobstructed during design flood	0.30
All shapes, partially obstructed during design flood	0.20

0.007 = coefficient (in.^{2}/h/ft^3) corresponding to a factor of safety of 1.0;

R = worst case rate of rise and fall (ft/h);

A_e = total enclosed area (ft^2); and

c = opening coefficient of discharge given in Table 2-2.

3.0 HIGH RISK FLOOD HAZARD AREAS

3.1 SCOPE

The requirements of Section 3 shall apply to new construction in high risk flood hazard areas subject to one or more of the following hazards: alluvial fan flooding, flash floods, mudslides, erosion, high velocity flows, high velocity wave action, and damage-causing ice or debris.

3.2 ALLUVIAL FAN AREAS

Construction of structures shall be prohibited at the apex of an alluvial fan, and in the fan's meandering flow paths. Construction in other areas of the alluvial fan shall meet the following requirements:

1. The elevation of the lowest floor shall be a minimum of 1 ft above the highest adjacent grade, or higher, if required on a community's flood hazard map.
2. Foundations shall be designed and constructed to resist scour caused by the actual flow velocity but not less than 5 ft/s. Determination of actual flow velocities shall be based on a review of a community's flood hazard map and flood hazard study, or on hydraulic calculations.
3. Design and construction shall resist all load combinations specified in Section 1.4.

3.2.1 Protective Works in Alluvial Fan Areas

Structures shall be prohibited in alluvial fan areas unless: 1) a whole alluvial fan flood damage reduction project has been designed and constructed to safely pass the design flood at the apex, within the capacity of the constructed channel(s); 2) design and construction do not divert flood flows and debris toward other structures, nor increase flood velocities and depths elsewhere on the alluvial fan; and 3) such construction satisfies the requirements of Section 1.5.2 and a maintenance and operations plan for the protective works is provided.

3.3 FLASH FLOOD AREAS

Structures shall not be constructed in areas subject to flash flooding equal to or less than design flood conditions.

Areas suspected of being subject to flash floods shall be investigated to obtain historical information on past events. The investigation shall also include analysis of historic rainfall and runoff data for the watershed. Results of such analyses shall be documented in an engineering report, which defines the methodology and data used to conclude whether the area in question has the potential for flash flooding.

3.3.1 Protective Works in Flash Flood Areas

Structures shall not be permitted in areas subject to flash floods, unless protective works have been determined to provide protection during the design flood event, where such construction satisfies the requirements of Section 1.5.2, and where a maintenance and operations plan for the protective works has been provided.

3.4 MUDSLIDE AREAS

Structures shall not be constructed in areas subject to mudslides during periods of rainfall and runoff. Areas suspected of being subject to inundation by mudslides shall be investigated to obtain historical information on past flood events. The investigation shall also include analysis of the source area for potential overland or channel erosion, bank failure, hill-slope failure, and rainfall/runoff potential. Results of such analyses shall be documented in an engineering report, which defines the methodology and data used to conclude whether the area in question has potential for future mudslides.

3.4.1 Protective Works in Mudslide Areas

Structures shall not be constructed in areas subject to mudslides, unless protective works have been determined to provide protection during the design flood event, such construction satisfies the requirements of Section 1.5.2, and a maintenance and operations plan for the protective works has been provided.

3.5 EROSION PRONE AREAS

Structures shall not be constructed within floodplains subject to erosion from such phenomenon as caving banks, meandering streams or eroding shorelines, where such erosion is predicted to affect the structure unless the structure is protected as specified in Section 3.5.1.

Erosion prone areas shall be determined by analyzing available studies, historical data, watershed trends, average annual erosion rates, wave effects, flood velocities and duration of flow, geotechnical data, and existing protective works. Results of these analyses shall be documented in an engineering report, which defines the data and methodology used to identify erosion prone areas.

3.5.1 Protective Works in Erosion Prone Areas

The limits of an erosion prone area shall be subject to revision where protective works have been designed and constructed to control erosion processes during all flow and wave conditions up to and including the design flood, and where a maintenance and operations plan for the protective works has been provided.

3.6 HIGH VELOCITY FLOW AREAS

High velocity flow areas shall be identified from a community's flood hazard map or flood hazard study, or from hydraulic analyses. The results of such analyses shall be documented in an engineering report, which defines the methodology and data used to conclude whether a site is susceptible to high velocity flows.

3.6.1 Protective Works in High Velocity Flow Areas

Structures shall not be constructed in high velocity flow areas, unless protective works have been determined to provide protection during the design flood event, such construction satisfies the requirements of Section 1.5.2, and a maintenance and operations plan for the protective works has been provided.

3.7 HIGH VELOCITY WAVE ACTION AREAS

Structures shall not be constructed in areas susceptible to high velocity wave action, unless the design and construction meet the requirements of Section 4.

3.8 ICEJAM AND DEBRIS AREAS

Structures shall not be constructed within floodplains that are subject to transportation of damage-causing ice or debris during floods up to and including the design flood.

The potential for ice or debris capable of inducing or causing loads exceeding design loads shall be identified from a community's flood hazard map or flood hazard study, or from hydraulic and other analyses. The results of such analyses shall be documented in an engineering report, which defines the methodology and data used to conclude whether a site is susceptible to icejams and debris effects.

3.8.1 Protective Works in Icejam and Debris Areas

Structures in icejam and debris areas shall have protective works to provide protection during the design flood event and meet the requirements of Section 1.5.2. Maintenance and operations plan for the protective works shall be provided.

4.0 FLOOD HAZARD AREAS SUJBECT TO HIGH VELOCITY WAVE ACTION

4.1 SCOPE

The requirements of Section 4 shall apply to new construction in flood hazard areas subject to high velocity wave action.

4.1.1 Identification of Areas Subject to High Velocity Wave Action

For the purposes of this Standard, "subject to high velocity wave action" shall mean those locations where an area has been designated as subject to high velocity wave action on a community's flood hazard map, or:

1. where the stillwater depth of the design flood above the eroded ground is greater than or equal to 3.8 ft, i.e., sufficient to support a wave height equal to or greater than 3 ft (see Fig. 4-1), and where conditions are conducive to the formation and propagation of such waves, or
2. where the eroded ground elevation under design flood conditions is 3 ft or more below the maximum wave runup elevation (see Fig. 4-2).

4.2 GENERAL

Design for flood hazard areas subject to high velocity wave action shall include the following:

1. waves breaking against the side or underside of the structure
2. drag, inertia and other wave-induced forces acting on structural members supporting elevated structures
3. uplift forces from breaking waves striking the undersides of structures
4. wave runup forces including those deflected by the structure
5. erosion and scour

4.3 SITING

New construction within flood hazard areas subject to high velocity wave action shall:

1. be located landward of the reach of mean high tide.
2. be sited landward of shoreline construction setbacks.
3. not remove or otherwise alter sand dunes and mangrove stands, unless the alterations will not reduce the wave and flow dissipation characteristics of the sand dunes or mangrove stands.

4.4 ELEVATION REQUIREMENTS

The bottom of the lowest horizontal structural member of the lowest floor shall be at or above the DFE, in conformance with the requirements of Table 4-1. The actual required height above the DFE shall be determined by the structure category and the orientation of the lowest horizontal structural member relative to the direction of wave approach. Piles, pile caps, footings, mat or raft foundations, grade beams, columns and shear walls designed and constructed in accordance with Section 4.5 shall not be required to meet the elevation requirements of Table 4-1.

4.5 FOUNDATION REQUIREMENTS

4.5.1 General

Foundation systems located in flood hazard areas subject to high velocity wave action shall be designed to minimize forces acting on that system. Foundation systems shall be free of obstructions and attachments that will transfer flood forces to the structural system, or that will restrict or eliminate free passage of high velocity flood waters and waves during design flood conditions.

New construction shall be supported on piles or columns. Mat or raft foundations shall not be used unless the top of the mat or raft foundation is below the eroded ground elevation. Columns shall be connected to and extend upward from the mat or raft foundation to a point at or above the DFE, as required by Table 4-1.

Where surface or sub-surface conditions consist of non-erodible soil which prevents the use of pile or deeply embedded column foundations, spread footings or mat foundations shall be permitted provided they are anchored, if necessary to prevent sliding, uplift or overturning, to non-erodible soil with sufficient strength to withstand forces from the combination of loads in Section 1.4.2.

Shear walls below the DFE shall comply with the following:

1. all construction, including the shear walls, other foundation elements, structural frame and connections, shall be designed for applicable loads,
2. shear walls oriented perpendicular to the direction of flood flow and wave approach shall be staggered so as not to form a continuous shear wall or an enclosed area below the DFE. An unobstructed area equal to one-half the area blocked by a shear wall perpendicular to the direction of flood flow and wave approach shall be provided adjacent to each shear wall.

FLOOD RESISTANT DESIGN AND CONSTRUCTION

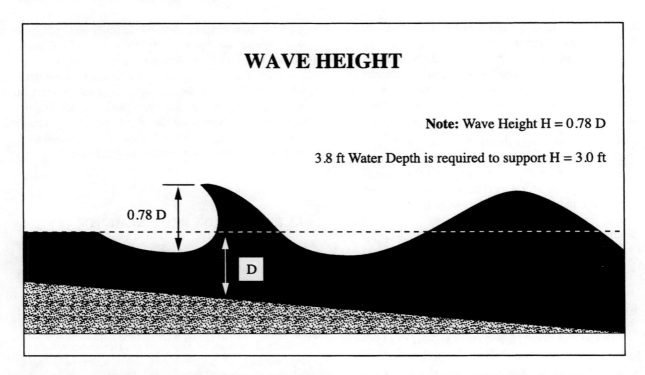

FIGURE 4-1. Definition Sketch for Still-Water Depth and Wave Height [C18]. (Reproduced with Permission. Copyright 1977 National Academy of Sciences.)

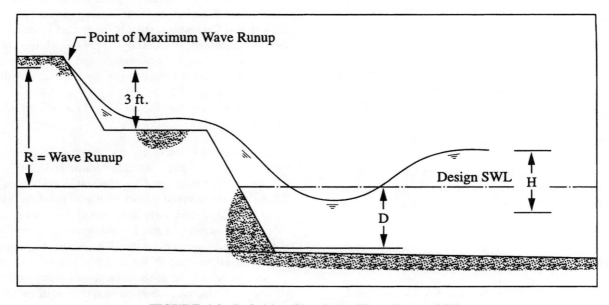

FIGURE 4-2. Definition Sketch for Wave Runup [C7].

TABLE 4-1. Minimum Elevation of Bottom of Lowest Supporting Horizontal Structural Member of Lowest Floor Relative to Base Flood Elevation (BFE) or Design Flood Elevation (DFE)—Flood Hazard Areas Subject to High Velocity Wave Action

Structure Category[a]	Member Orientation Relative to the Direction of Wave Approach	
	Parallel[b]	Perpendicular[b]
I	at DFE	at DFE
II	at DFE	at DFE
III	at BFE + 1 ft or DFE, whichever is higher	at BFE + 1 ft or DFE, whichever is higher
IV	at BFE + 1 ft or DFE, whichever is higher	at BFE + 2 ft or DFE, whichever is higher

[a]See Table 1-1 for Structure descriptions.
[b]Orientation of lowest horizontal structural member relative to the general direction of wave approach: parallel shall mean less than or equal to ±20° from the direction of approach; perpendicular shall mean greater than ±20° from the direction of approach.

4.5.2 Special Geotechnical Considerations

In addition to the requirements of Section 2.5.1, foundation design shall account for instability and decreased structural capacity associated with erosion due to wind, waves, currents, local scour, storm induced erosion, and shoreline movement.

4.5.3 Foundation Depth

The foundation shall extend to a depth sufficient to provide the support described in Section 2.5, taking into account the erosion and scour of the supporting soil during design flood—shoreline movement and lesser—conditions, as predicted by an erosion analysis.

4.5.4 Use of Fill

The use of fill material for structural support or protection shall be prohibited in flood hazard areas subject to high velocity wave action. Use of nonstructural fill beneath a structure is prohibited, except where used for minimal site grading to meet local drainage requirements, or placed outside the structure footprint for landscaping and site grading. All site grading and fill placement shall be performed so as to assure the free passage of water beneath the structure, and to avoid the diversion or ramping of water and wave runup toward any structure.

4.5.5 Pile Foundations

Except as provided for under Section 4.5.1, all foundations constructed in erodible soils shall be founded on piles. Piles that are jetted shall be seated by driving. Piles and connections to piles shall be designed in accordance with Section 5.

In erodible soils, pile tip penetration shall be to a minimum depth of 10 ft below mean water level (−10 ft MWL), or to refusal, whichever is shallower. Solely for the purpose of determining the ability to penetrate to −10 ft MWL, exclusive of providing lateral stability against wind, water, wave and other loads and exclusive of bearing capacity requirements, refusal shall be as defined by the geotechnical investigation. In the event that changed conditions are encountered during construction and/or refusal is not reached during pile installation, additional geotechnical investigations and a revised pile design shall be completed.

For the calculation of pile reaction and bearing capacity in erodible soils, the design shall consider that local scour and liquefaction of the soil during design flood conditions will render as non-supportive at least 5 ft of soil below the ground elevation at the point of pile penetration. This non-supportive soil shall not be considered in the design. Calculation of local scour effects during design flood conditions shall include the interactive effects of pilings or other foundation elements in close proximity to one another.

Piles, with attached floodproofed enclosures in excess of 2 ft in any dimension, and conforming to the limitations established in Sections 7.2.1 and 7.3.1, shall be designed to withstand the additional loading introduced by the enclosure. Scour around the base of a larger permanent enclosure will also be increased and pile design shall call for a deeper pile embedment.

4.5.5.1 Piles Terminating in Caps at or Below Grade

Foundations comprised of a number of single piles or pile clusters terminating in reinforced concrete pile caps at or below eroded ground elevation, shall be designed for the combination of loads in Section 1.4.2. Individual pile caps shall be laterally connected by a system of reinforced concrete grade beams. Embedment of the pile into the pile cap shall be designed in accordance with Section 5.

The pile cap shall be designed and constructed to be structurally stable, without relying on supporting soil around or under the cap. Pile-to-pile cap and pile cap-to-column connections shall be designed to withstand expected hydrodynamic forces including wave and debris impact.

4.5.5.2 Piles Extending to Superstructure (Structure Framing)

The portion of a pile that extends above the eroded ground elevation to elevate a structure, shall be designed as a column. Pile spacing shall take into account the design bearing capacity, uplift and overturning resistance.

Bracing shall be required where the exposed pile length, after accounting for scour and erosion, exceeds 10 times the mean least pile dimension normal to its length. Bracing shall be designed in accordance with Sections 4.5.8 and 5.

4.5.5.3 Wood Piles

Wood piles shall be preservative-treated in accordance with Section 6. Round wooden piles shall have a minimum diameter of 8 in. when measured 3 ft from the butt, and 8 in. minimum diameter at the tip. The minimum size for square piles shall be 8 in. on a side where predicted erosion is expected to expose pile lengths equal to or less than 12 ft and, shall be 10 in. on a side where the predicted erosion is expected to expose pile lengths greater than 12 ft.

Wood piles which are directly connected to beams supporting an elevated lowest floor shall be individually secured thereto by means of at least two hot-dipped galvanized or stainless steel bolts, nuts and washers, of sufficient size and number to resist the forces resulting from the combinations of loads in Section 1.4.2.

4.5.5.4 Steel H Piles

Steel piles of rolled HP-Sections or built-up sections shall consist of a corrosion resistant material or be protected from corrosion, by a corrosion resistant coating or by cathodic protection, in accordance with Section 6. Built-up sections shall have a web thickness equal to the flange thickness; the web and flange shall be continuously welded together.

4.5.5.5 Concrete-Filled Steel Pipe Piles and Shells

Steel components of concrete-filled steel pipe piles and shells shall be protected with a corrosion resistant coating, in accordance with Section 6. Concrete and steel pipe piles or shells shall be designed in accordance with Section 5.

4.5.5.6 Prestressed Concrete Piles

Prestressed concrete piles shall be designed, manufactured and installed in accordance with Sections 5 and 6.

4.5.6 Columns

Columns, including wood posts, reinforced masonry columns, reinforced concrete columns, and associated connections shall be designed and constructed to resist wind, water, wave and other flood-related forces, in accordance with Section 5. Column spacing shall take into account the design bearing capacity, uplift and overturning resistance but shall be spaced not less than 8 ft center to center. Where founded on erodible soils, column supports shall extend to at least -10 ft MWL, or shall be supported by and anchored to a pile, mat or raft foundation meeting the requirements of Section 4.5.1.

4.5.6.1 Wood Posts

Wood posts shall be 10 in. by 10 in. minimum for square posts or 10 in. in diameter for round posts. Wood posts shall be preservative-treated in conformance with Section 6. Where founded on non-erodible soil, posts shall be adequately embedded and anchored to the footing which shall be anchored to the substrata to prevent pullout during design flood conditions.

4.5.6.2 Reinforced Masonry Columns

Reinforced masonry columns shall comply with ACI 530/ASCE 5/TMS 402 [3].

4.5.6.3 Reinforced Concrete Columns

Concrete columns shall comply with ACI 318 [4].

4.5.7 Grade Beams

Grade beams shall be constructed with their upper surface at or below natural grade, shall be structurally connected to the primary foundation system to provide additional lateral support, and shall be structurally independent of decks, patios and concrete pads. Grade beams shall be designed to perform their structural function without the aid of supporting earth and while exposed to flood forces.

4.5.8 Bracing

Bracing shall be permitted when more than 8 ft of the column is exposed above the eroded ground elevation. Bracing and its connections to the primary vertical structural members shall be designed to withstand the lateral force of waves and debris impact in accordance with Section 5. The braces shall be designed to resist forces acting in both the plane of brace resistance and perpendicular to the plane of brace resistance. The foundation system shall be designed to account for the additional loads transferred from the bracing to the main supporting structures. Cross-bracing perpendicular to the primary direction of wave and hydrodynamic forces shall be restricted to tensile bracing using steel rods or steel cables. Cross-bracing parallel to the primary direction of wave and hydrodynamic forces shall not be restricted.

Steel rods used for cross-bracing shall be galvanized or of non-corrosive material with a minimum diameter of 1/2 in. An equivalent system of corrosion-resistant high tensile steel cables and turnbuckles may be used in lieu of solid rods. Where rods or cables are used for bracing, they shall be tied together with a clamp at the center cross point.

The smallest nominal dimension of any wood member used for cross-bracing shall be not less than 4 in.

Cross-bracing shall be attached to the main vertical structural elements with bolts, nuts and plate washers. Connections to the main vertical structural elements shall be within 12 in. of the lowest elevated floor support beams, and as near to grade as possible.

Knee braces are permitted in all directions relative to flood forces, and shall extend not more than 3 ft below the DFE.

4.6 ENCLOSED AREAS BELOW DESIGN FLOOD ELEVATION

Enclosed areas below the DFE shall only be allowed where all of the following conditions are met:

1. enclosures shall be designed and constructed as breakaway construction that will not adversely affect any structure by redirecting floodwaters or by producing debris capable of causing damage to structures

2. enclosed areas shall be used solely for parking, building access and storage

4.6.1 Breakaway Walls

Breakaway walls and other similar non-load bearing components, including open-wood lattice work and insect screening, shall be designed and constructed to fail under design flood or lesser conditions, without imparting additional flood loads to the foundation or superstructure, and without producing debris damage to the structure or adjacent structures. Breakaway walls and their connections shall be designed in accordance with the requirements of Section 5.3.2.2 of ASCE 7 (Standard Reference [1]). Utility lines shall not be attached to or pass through or be included in breakaway construction.

4.7 EROSION CONTROL STRUCTURES

Bulkheads, seawalls, revetments and other erosion control structures shall not be connected to the foundation or superstructure of a structure. Bulkheads, seawalls, revetments and other erosion control structures shall be designed and constructed so as to not focus or increase flood forces or erosion impacts on the foundation or superstructure of any structure.

4.8 DECKS, CONCRETE PADS, AND PATIOS

Decks, concrete pads and patios below the DFE shall be permitted beneath or adjacent to a structure, provided they are structurally independent from the primary structural foundation system of the structure, and provided they do not adversely affect adjacent structures by redirecting floodwaters or becoming sources of debris.

Decks, concrete pads and patios shall not transfer flood loads to the main structure, shall be constructed to break away cleanly during design flood conditions, shall be friable, and shall not produce debris capable of causing significant damage to any structure.

Reinforcing, including welded wire fabric, shall not be used in order to minimize the potential for concrete pads being a source of debris. Pad thickness shall not exceed 4 in.

Decks and patios subject to floating shall be adequately anchored to prevent flotation and meet all the provisions of Section 4.5. Other than minor regrading for proper drainage, no fill shall be used to elevate decks, concrete pads, or patios above natural grade.

FLOOD RESISTANT DESIGN AND CONSTRUCTION

Permanent benches or railings capable of obstructing flow, shall be prohibited.

5.0 DESIGN

5.1 GENERAL

The main vertical and lateral force resisting structural systems of new construction located in flood hazard areas, shall be designed and constructed in accordance with the requirements of this Section.

5.2 VERTICAL STRUCTURAL SYSTEMS

5.2.1 Masonry Walls

Masonry walls shall be designed in accordance with the requirements of ACI 530/ASCE 5/TMS 402 *Building Code Requirements for Masonry Structures* [3].

5.2.2 Concrete Walls

Design of concrete foundation walls shall be in accordance with ACI 318 *Building Code Requirements for Structural Concrete* [4].

5.2.3 Lateral Resistance of Open Foundation Systems

Bracing used for lateral support of foundation systems in flood hazard areas shall not exceed those limitations imposed in Sections 2–4.

5.2.4 Piles

5.2.4.1 General

The types of piles selected for any given installation shall be based on the geotechnical characteristics of the soil as required by Section 2.5.1.

5.2.4.2 Timber Piles

Conditioning and preservative treatments of timber piles shall be in accordance with Section 6.2.5. Considerations shall be given to the use of pile fittings at the butt, tip, and at designated intervals along the pile length for the protection of piles during installation. Round timber piles shall conform to ASTM D25 *Standard Specification for Round Timber Piles* [5].

5.2.4.3 Steel HP-Section Piles

Piles shall be rolled HP-sections, or built-up sections with the web thickness equal to the flange thickness, and with the web and flanges continuously welded together.

The thickness of metal shall be 0.40 in. minimum. The thickness shall also be based on the loss of section due to corrosion, unless corrosion protection is provided in the form of concrete, bituminous or plastic (epoxy) coatings, or cathodic protection. Damage to coatings during installation shall be avoided, and if damage should occur repairs shall be made in accordance with manufacturers' recommendations and applicable codes, standards, and regulations.

Pile tip reinforcing, splicing, fittings, and cap plates shall be provided, as required.

5.2.4.4 Concrete-Filled Steel Pipe Piles and Shells

Pipe for concrete-filled steel pipe piles shall conform to ASTM-A252 *Specification for Welded and Steel Pipe Piles* [6]. Pile tip reinforcing, splicing, fittings, and cap plates shall be specified by the designer, as required. Flat steel plates closing the tip of pipe piles shall be of a diameter not greater than 3/4 in. more than the outside diameter of the pipe.

Concrete and aggregate materials for concrete to be cast shall be specified with consideration given to the difficulty of placement conditions.

5.2.4.5 Precast (Including Prestressed) Concrete Piles

Pile dimension shall be 10 in. minimum for piles of uniform section and 8 in. minimum for tapered piles.

For piles subject to exposure from brackish water, seawater or spray from these sources, cover for reinforcement shall be not less than 3 in. for precast piles and not less than 2 1/2 in. for precast, prestressed piles.

5.2.4.6 Cast-in-Place Concrete Piles

Concrete and steel reinforcement used in cast-in-place concrete piles shall conform to the requirements of ACI 318 *Building Code Requirements for Structural Concrete* [4].

5.2.4.7 Pile Capacity

Piles shall be designed to carry the loads imposed by the combinations of loads of Section 1.4.2, and to withstand installation forces.

Unless exceeded by moments indicated by analysis of applied loads, the piles shall be designed for a minimum moment produced by an eccentricity of 0.10 times the equivalent diameter of the pile, times the axial load of the pile.

The minimum required lateral resistance of an individual pile shall be at least 5% of the axial load on the pile.

5.2.4.8 Capacity of the Supporting Soils

Soil values pertaining to friction, end bearing resistance, and settlement of single piles and pile groups shall be based on the geotechnical characteristics of the soil as required by Section 2.5.1.

For piles spaced more than three pile diameters center-to-center, the diameter of the soil that shall be assumed reacting laterally on each pile shall have a maximum equivalent diameter equal to three times the diameter of the pile.

5.2.4.9 Minimum Penetration

Pile penetration into acceptable bearing strata shall be a minimum depth sufficient to allow distribution of the pile load to the supporting soils, including a consideration for reduction in soils due to the effects of scour and erosion, in accordance with Section 2.5.2 and/or Section 4.5.3.

If the minimum penetration is less than 10 ft into non-erodible soil, or less than 20 ft into erodible soil, special provisions such as hardened tips driven into the refusing stratum, drilled sockets, or drilled dowels shall be made to secure the tips of the pile against lateral displacement and vertical movement.

5.2.4.10 Foundation Pile Spacing

The spacing of foundation piles shall provide adequate distribution of the load on a pile group to the supporting soil. Minimum pile spacing within the group shall be three times the diameter of the pile.

5.2.4.11 Pile Caps

For pile caps in contact with the ground, piles shall be designed to carry the total superimposed vertical load from the structure above with no allowance made for the supporting value of the soil under the pile caps.

5.2.4.12 Timber Pile Connections

In timber piles, bolts for cap or beam-to-pile connections shall be 5/8 in. in diameter minimum. Bolt holes shall be staggered with a maximum diameter of 1/16 in. greater than the bolt diameter. The dimension from the edge of holes to the pile or beam edge shall be 2 in. minimum. Notching of pile tops shall not exceed 50% of the pile diameter. Other pile-to-beam connections are acceptable provided they are demonstrably equal to or superior for the applications, and do not depend upon nailing for attachment of individual members.

5.2.4.13 Timber Piles Not in Tension

For timber piles not in tension, and connected to timber caps, the tops of the piles shall be secured to the caps with spiral-drive drift bolts, metal plates, or bolted timber scabs.

For timber piles not in tension, and connected to concrete caps, the tops of the piles shall have a minimum 4 in. embedment into the concrete pile caps.

5.2.4.14 Timber Piles in Tension

For timber piles in tension, piles shall be embedded into pile caps a minimum of 6 in., with a connection designed for tension made to pile caps. Connections to timber caps shall be made with timber scabs or metal straps, and headed shear bolts.

Connections to concrete caps shall have the tops of the piles embedded to satisfy requirements of shear stress in the timber (parallel to grain) and shear stress in the concrete. Embedments shall satisfy the requirements of Sections 2–4. Connections shall be made with metal straps, headed shear bolts, or other forms of positive tension resisting devices that develop the necessary shear in the concrete without causing failure of the wood.

5.2.4.15 Steel Piles Not in Tension

For steel piles not in tension, and where structural design of piles of significant length depend on bending in the piles for stability, the tops of steel piles shall be tied into concrete caps with reinforcing steel or structural sections welded to the pile, and lapped with the cap reinforcement. The minimum lap shall be 6 in.

5.2.4.16 Steel Piles in Tension

For steel piles in tension, the tops of steel piles shall be tied into concrete caps with reinforcing steel or structural sections welded to the pile, and lapped with the cap reinforcement. The minimum lap shall be 6 in. Bond stress between concrete and steel shall not exceed $0.02 f'_c$, where f'_c is the specified compressive strength of concrete.

5.2.4.17 Concrete Piles Not in Tension

For concrete piles not in tension, the tops of the piles shall have a minimum 3 in. embedment into the pile cap.

5.2.4.18 Concrete Piles in Tension

For concrete piles in tension, the tops of concrete piles in concrete caps shall be adequately doweled and embedded into the cap to resist tension loads. The tops of piles shall have a minimum 3 in. embedment into the pile cap.

5.2.4.19 Pile Splicing

Splices shall be constructed to provide and maintain the alignment and position of the component parts of the pile during installation and subsequent thereto. Splices shall be of adequate strength to transmit the axial and lateral loads, and the moments at the section involved.

Pile splices that cannot be visually inspected after pile installation shall develop the greater of at least 50% of the capacity of the pile, or the moment, shear, and tension that would result from an assumed eccentricity of the resultant pile load of 3 in.

In areas subject to high-velocity wave action and subject to flood-related erosion, timber pile splicing shall be made below the scour elevation, as determined by an erosion analysis.

5.2.4.20 Mixed Types of Piling and Multiple Types of Installation Methodology

The use of mixed types or capacities of piling and different types of installation equipment or methods shall consider, in addition to the relative lateral load capacities, an analysis of the additional effects on the superstructure of differential elastic shortening and settlement.

5.2.5 Posts, Piers, and Columns

5.2.5.1 Application

Posts, piers, and columns shall resist the effects of the combinations of loads in Section 1.4.2.

5.2.5.2 Reinforced Concrete and Masonry Columns

Reinforced concrete columns shall comply with ACI 318 *Building Code Requirements for Structural Concrete* [4].

Masonry columns shall comply with ACI 530/ASCE 5/TMS 402 [3].

5.3 FOOTINGS

The tops of footings shall not be higher than the eroded ground elevation.

5.4 MATS, RAFTS, AND SLABS

Where permitted in accordance with Sections 2 and 4, mats, rafts, and other concrete slabs below the elevations specified in Tables 2-1 and 4-1 shall be designed as reinforced structural slabs, supported directly on the soil or on grade beams that are in turn supported on the soil. The design shall provide for the effects of additional lateral loads from flood forces transferred to the structure through these elements.

5.5 GRADE BEAMS

Design of grade beams and other foundation elements shall provide for the effects of lateral bracing offered by grade beams on the structure.

Grade beams, either of wood or reinforced concrete, shall be firmly attached to vertical members to transfer all vertical and lateral forces acting on the grade beam as a result of flooding, including wave action and the effects of debris, scour and erosion. The design of the grade beam members shall include bi-axial beam action to support the vertical and lateral forces acting the full length of members undermined by scour and erosion.

5.6 ANCHORAGE/CONNECTIONS

The structure shall be designed to resist effects of vertical loads, including uplift, and lateral loads due to the load combinations specified in Section 1.4.2. Stringers or beams shall be attached to the substructure or directly to piles, piers, and walls with bolted or welded connections.

Washers shall be used under all nuts and bolt heads bearing directly on wood. Notches in timber posts and piles shall not exceed 50% of the cross-section of the post or pile. All nuts, bolts and washers shall be corrosion-resistant.

Adequate anchorage shall also be provided for storage tanks, sealed conduits and pipes, lined pits, sumps and all other similar structures which have a negligible weight of their own or are subject to flotation or lateral movement during the design flood.

6.0 MATERIALS

6.1 GENERAL

New construction in flood hazard areas shall be constructed with flood-damage-resistant materials.

Exposed structural and non-structural construction materials, including connections, shall be capable of resisting damage, deterioration, corrosion, or decay due to precipitation, wind driven water, salt spray or other corrosive agents known to be present.

Structural and non-structural construction materials, including connections, below the elevations specified in Table 6-1 shall be capable of resisting damage, deterioration, corrosion or decay due to direct and prolonged contact with floodwaters, associated with design flood conditions.

Materials used in new construction in flood hazard areas shall have sufficient strength, rigidity, and durability to adequately resist all flood-related and other loads during installation.

6.2 SPECIFIC MATERIALS REQUIREMENTS FOR FLOOD HAZARD AREAS

6.2.1 Metal Connectors and Fasteners

Metal plates, connectors, screws, bolts, nails, and other fasteners exposed to direct contact by flood water, precipitation, or wind-driven water shall be hot-dip galvanized after manufacture or fabrication, in accordance with: ASTM A-123 *Specifications for Zinc (Hot-Galvanized) Coatings on Iron and Steel Products* [7], ASTM A-153 *Specifications for Zinc Coating (Hot-Dip) on Iron and Steel Hardware* [8] or ASTM A-525 *Specifications for General Requirements for Steel Sheet, Zinc Coated (Galvanized) by the Hot-Dip Process* [9].

6.2.2 Structural Steel

H-Section Piles shall conform to ASTM-A36/A36M *Specification for Structural Steel* [10], ASTM-A572/A572M *Specification for High-Strength Low-Alloy Columbium-Vanadium Steels of Structural Quality* [11], or ASTM A690/A690M *Specification for High-Strength Low-Alloy Steel H-piles and Sheet Piling for Use in Marine Environments* [12].

6.2.2.1 Corrosive Environments

Structural steel exposed to direct contact with salt water, salt spray or other corrosive agents known to be present shall be hot-dipped galvanized after fabrication. Secondary components such as angles, bars, straps, and anchoring devices shall be stainless steel or hot-dipped galvanized after fabrication, in accordance with Section 6.2.1.

6.2.2.2. Non-Corrosive Environments

In areas where salt spray and other corrosive agents are not known to be present, exposed structural steel shall either meet the requirements of Section 6.2.2.1, or be primed, coated, plated, or other-

TABLE 6-1. Minimum Elevation, Relative to Base Flood Elevation (BFE) or Design Flood Elevation (DFE), Below Which Flood Damage Resistant Materials Shall be Used

	Use Flood Damage Resistant Materials Below Elevation		
		Areas Subject to High Velocity Wave Action	
Structure Category[a]	Areas Not Subject to High Velocity Wave Action	Orientation Parallel[b]	Orientation Perpendicular[b]
I	at DFE	at DFE	at DFE
II	at DFE	at BFE + 1 ft or DFE, whichever is higher	at BFE + 2 ft or DFE, whichever is higher
III	at DFE	at BFE + 2 ft or DFE, whichever is higher	at BFE + 3 ft or DFE, whichever is higher
IV	at BFE + 1 ft or DFE, whichever is higher	at BFE + 2 ft or DFE, whichever is higher	at BFE + 3 ft or DFE, whichever is higher

[a]See Table 1-1 for Structure descriptions.
[b]Orientation of lowest horizontal structural member relative to the general direction of wave approach: parallel shall mean less than or equal to ±20° from the direction of approach; perpendicular shall mean greater than ±20° from the direction of approach.

wise protected against corrosion due to direct contact with floodwaters, precipitation or wind-driven water.

Secondary components such as angles, bars, straps, and anchoring devices shall be stainless steel or hot-dipped galvanized after fabrication, in accordance with Section 6.2.1.

Damage to protective finishes and coatings caused by handling or installation shall be repaired using procedures that result in protection equivalent to the requirements stated above.

6.2.3 Concrete

Ingredients of concrete, including admixtures and reinforcing steel; quality of concrete; and the design and construction thereof shall comply with ACI 318 *Building Code Requirements for Structural Concrete* [4]

6.2.4 Masonry

Materials used in masonry construction, including masonry units, mortar, grout, reinforcing steel and accessories; quality of masonry; and the design and construction thereof shall comply with ACI 530/ASCE 5/TMS 402 *Buildings Code Requirements for Masonry Structures* [3].

6.2.5 Wood and Timber

Wood members, exposed or enclosed, solid or built-up, shall be naturally decay resistant or pressure treated with preservatives to resist damage, deterioration or decay due to: insect infestation, dry rot or fungi; contact with floodwaters, precipitation, wind driven water, salt spray or other corrosive agents known to be present.

6.2.6 Finishes

Interior wall finishes and trim shall be flood-damage-resistant material.

7.0 DRY AND WET FLOODPROOFING

7.1 SCOPE

This section addresses design and construction requirements for floodproofing new construction in flood hazard areas. Design and construction of floodproofing shall conform with this Standard and with Section 5 of ASCE-7 [1].

Design of floodproofing measures shall account for: nature of flood-related hazards; frequency, depth and duration of flooding; rate of floodwater rise and fall; floodwater temperature; soil characteristics; flood-borne contaminants and debris; flood warning time; access to and from floodproofed areas; structure occupancy and use; and functional dependence.

7.2 DRY FLOODPROOFING

Dry floodproofing shall be accomplished through the use of flood-damage-resistant materials and techniques that render the portions of a structure substantially impermeable to the passage of floodwater below the elevations specified in Table 7-1. Sump pumps shall be provided to remove water accumulated due to any passage of vapor and seepage of water during the flooding event. Sump pumps shall not be relied upon as a means of dry floodproofing. All materials in contact with floodwaters shall conform with the requirements of Section 6.

7.2.1 Dry Floodproofing Restrictions

Dry floodproofing of non-residential structures and non-residential areas of mixed use structures shall not be permitted unless they are located outside of high risk flood hazard areas. Dry floodproofing of residential structures or residential areas of mixed use structures shall not be permitted.

Dry floodproofing shall be limited to the following:

1. where flood velocities adjacent to the structure are less than or equal to 5 ft/s during the design flood; and
2. if human intervention is required to activate or implement dry floodproofing, the flood warning

TABLE 7-1. Minimum Elevation of Floodproofing, Relative to Base Flood Elevation (BFE) or Design Flood Elevation (DFE)—Outside of High Risk Flood Hazard Areas[a]

Structure Category[b]	Minimum Elevation of Floodproofing[c]
I	at BFE + 1 ft or DFE, whichever is higher
II	at BFE + 1 ft or DFE, whichever is higher
III	at BFE + 1 ft or DFE, whichever is higher
IV	at BFE + 2 ft or DFE, whichever is higher

[a]See Sections 7.2 and 7.3.1
[b]See Table 1-1 for structure descriptions.
[c]Wet or dry floodproofing shall extend to the same level.

time (alerting potential flood victims of pending flood situation) shall be a minimum of 12 h, unless the community operates a flood warning system and implements an emergency plan to ensure safe evacuation of flood hazard areas, in which case the minimum flood warning time shall be 2 h, or be based on the following criteria, whichever results in a longer warning time:

a) time to notify person(s) responsible for installation of floodproofing measures, plus;
b) time for responsible persons to travel to structure to be floodproofed, plus;
c) time to install floodproofing measures, plus;
d) time to evacuate all persons out of the floodplain.

7.2.2 Dry Floodproofing Requirements

Dry floodproofed areas of structures shall:

1. be designed and constructed so that any area below the floodproofed design level, together with attendant utilities and sanitary facilities, is flood-resistant with walls that are substantially impermeable to the passage of water. Basement walls, floors, and flood shields shall be built with the capacity to resist hydrostatic, hydrodynamic and other flood-related loads, including the effects of buoyancy resulting from flooding to the elevation listed in Table 7-1; and
2. have any soil or fill on the sides of the structure compacted and protected against erosion and scour in accordance with Section 2.5.3.
3. have openings satisfying building code requirements, above the elevations specified in Tables 2-1 and 4-1, and capable of providing human ingress and egress during the design flood.

7.3 WET FLOODPROOFING

Wet floodproofing shall be accomplished through the use of flood-damage-resistant materials and techniques that minimize damage to a structure during periods where the lower portion of the structure is inundated by floodwater. All materials in contact with floodwaters shall conform with the requirements of Section 6.

7.3.1 Wet Floodproofing Restrictions

Wet floodproofing shall be allowed for Category 1 structures. For all other categories of structures, only in enclosed areas below the elevations listed in Table 7-1 used solely for parking, building access or storage, when such structures are located outside high risk flood hazard areas. For elevated structures located in high risk flood hazard areas, wet floodproofing shall be permitted only for the enclosed areas below the elevations listed in Table 7-1, used solely for parking, building access, or storage.

7.4 ACTIVE FLOODPROOFING

Active floodproofing requires human intervention prior to or during a flood, and shall be permitted only when all of the following conditions are satisfied:

1. all removable shields or covers for openings such as windows, doors and other wall penetrations shall be designed to resist flood loads specified in Section 1.4 of this Standard; and
2. where removable shields are to be used, a flood emergency plan approved by the authority having jurisdiction shall be provided which specifies: storage location(s) of the shields, the method of installation, conditions activating installation, and maintenance of shields and attachment devices. The emergency plan shall be permanently posted in at least two conspicuous locations within the structure. The emergency plan shall include provisions for periodic practice of installing shields, testing sump pumps, and inspecting necessary material and equipment to complete active floodproofing.

Active floodproofing shall not be permitted where flash floods occur or where floodwaters rise quickly, resulting in warning times less than that stipulated by Section 7.2.1.

8.0 UTILITIES

8.1 GENERAL

Utilities and attendant equipment shall not be located below the elevation specified in Table 8-1, unless permitted in Sections 8.2 through 8.4 and they are designed, constructed and installed to prevent floodwaters, including any back flow through the system, from entering or accumulating within the utility, mechanical components, and the areas of structure that are waterproofed or above DFE. Utilities shall not be mounted on or located along breakaway walls.

TABLE 8-1. Minimum Elevation of Utilities and Attendant Equipment Relative to Base Flood Elevation (BFE) or Design Flood Elevation (DFE)

Structure Category[a]	Locate Utilities and Attendant Equipment Above[b]		
	Areas Not Subject to High Velocity Wave Action	Areas Subject to High Velocity Wave Action	
		Orientation Parallel[c]	Orientation Perpendicular[c]
I	at DFE	at DFE	at DFE
II	at DFE	at BFE + 1 ft or DFE, whichever is higher	at BFE + 2 ft or DFE, whichever is higher
III	at BFE + 1 ft or DFE, whichever is higher	at BFE + 2 ft or DFE, whichever is higher	at BFE + 3 ft or DFE, whichever is higher
IV	at BFE + 1 ft or DFE, whichever is higher	at BFE + 2 ft or DFE, whichever is higher	at BFE + 3 ft or DFE, whichever is higher

[a] See Table 1-1 for Structure descriptions.
[b] Locate utilities and attendant equipment above elevations shown unless an exception is made in the text.
[c] Orientation of lowest horizontal structural member relative to the general direction of wave approach: parallel shall mean less than or equal to ±20° from the direction of approach; perpendicular shall mean greater than ±20° from the direction of approach.

Elevated exterior platforms for utilities and attendant equipment shall be supported on piles or columns, cantilevered, or knee braced to the main structure. If piles or columns are utilized, they shall be adequately embedded to account for erosion and local scour around the supports.

8.2 ELECTRICAL

8.2.1 Service Conduits and Cables

Electrical service conduits and cables below the DFE shall be waterproofed and buried to a depth sufficient to prevent movement, separation or loss due to erosion and scour under design flood conditions.

8.2.2 Exposed Conduits and Cables

Electrical supply lines emerging from underground shall be designed, constructed and installed to withstand flood-related loads, including the effects of buoyancy, hydrodynamic forces and debris impacts. Waterproofing or protective enclosures shall be provided for supply lines extending vertically to elevated structures. The enclosures shall be securely fastened to the structure; however, protective enclosures and electrical supply lines shall not be fastened to walls, enclosures or structures intended to break away under flood conditions, per Section 4.6.1. Electrical supply lines and protective enclosures installed below the elevations specified in Table 8-1 shall be sealed to prevent the entrance of floodwaters into electrical conduits and components.

8.2.3 Electric Meters

Electric meters shall be located above the elevation specified in Table 8-1 unless, the connection between the meter and electric lines extending vertically from the meter is within a waterproof enclosure.

8.2.4 Disconnect Switches and Circuit Breakers

The main disconnect switch and all circuit breakers shall be located above and be accessible from above the elevation specified in Table 8-1. Switches and circuit breakers shall be located no more than 6 1/2 ft above the floor, or platform installed to provide access to them.

8.2.5 Electric Elements Installed Below Minimum Elevations

Where electrical supply lines are located below the elevation specified in Table 8-1, they shall be installed so as to drain water away from panelboards, controllers, switches or other electrical equipment in accordance with NFPA 70 *National Electrical Code* [13].

Lighting circuits, switches, receptacles and lighting fixtures operating at a maximum voltage of 120 volts to ground shall be permitted below the elevation specified in Table 8-1, provided all electrical wiring, switches, receptacles and fixtures are suitable for continuous submergence in water, including the use of ground fault interruptable circuits, and shall contain no fibrous components. Where electrical wires are spliced below the elevation specified in Table 8-1, only submersible type splices shall be used.

All circuits, switches, receptacles, fixtures and other electrical components and equipment installed below the elevation specified in Table 8-1 shall be energized from a common distribution panel located above and accessible from above the elevation specified in Table 8-1.

In high risk flood hazard areas, main supply lines, meters and other exterior electrical components installed below the elevation specified in Table 8-1 shall be installed on a non-breakaway vertical structural element on the landward, down slope, or downstream side of the structure.

8.3 PLUMBING

For the purposes of this Standard, plumbing systems shall include: sanitary and rain runoff collection systems; sanitary facilities; water supply systems (including hot water heaters and water softeners); and rainwater runoff and sewage disposal systems.

8.3.1 Buried Plumbing Systems

Where installed underground, piping and plumbing systems providing service to a structure shall be buried to a depth sufficient to prevent movement, separation or loss due to erosion under design flood conditions.

8.3.2 Exposed Plumbing Systems

Plumbing systems and components emerging from underground shall be designed, constructed, installed and protected to withstand flood-related loads, including the effects of buoyancy, hydrodynamic forces and debris impacts.

8.3.3 Plumbing Systems Installed Below Minimum Elevations

Plumbing systems which have openings below the elevation specified in Table 8-1 shall be equipped with automatic backwater valves or other automatic backflow devices. Redundant devices requiring human intervention shall be permitted. Backwater valves or backflow devices shall be installed in each line that extends below the DFE. Plumbing systems shall be designed to prevent release of sewage into floodwaters and to prevent infiltration by floodwaters into the plumbing.

8.3.4 Sanitary Systems

Sanitary systems, including septic tanks, that must remain operational during the design flood or lesser floods, shall be equipped with a sealed holding tank and any necessary piping or equipment required to prevent sewage discharge during the design flood. The holding tank shall be vented to a point above the elevation specified in Table 8-1, and shall be sized to store at least 150% of the anticipated sewage flow during the design flood, and any accompanying period of saturated soil during which time sewage will not percolate. Holding tanks shall be designed and constructed to resist scour and erosion; holding tanks shall be designed, constructed, installed and anchored to resist 1.5 times the potential buoyant and other flood forces acting on an empty tank during design flood conditions.

8.4 MECHANICAL, HEATING, VENTILATION, AND AIR CONDITIONING

Fuel supply lines extending below the elevation specified in Table 8-1 shall be equipped with a float operated automatic control valve to shutoff fuel supply when floodwaters rise above the elevation of the supply line.

In flood hazard areas subject to high velocity wave action, and when located on the exterior of a structure, mechanical or HVAC equipment and systems shall be located on the landward side of the structure, above the elevation specified in Table 8-1.

Ductwork shall only be permitted below the elevation specified in Table 8-1, when it is designed, constructed and installed to resist all flood related loads, provided it (and ductwork insulation) is protected against damage, and provided it prevents floodwater entering or accumulating within the ductwork. Ductwork shall not be permitted to allow floodwaters to enter the structure.

8.4.1 Fuel Storage Tanks

Fuel storage tanks located below the elevation specified in Table 8-1 shall be designed, constructed, installed and anchored to resist all flood-related and other loads during the design flood, or lesser floods, without release of fuel into floodwaters or infiltration by floodwaters into the fuel.

Fuel storage tanks shall be designed, constructed, installed and anchored to resist 1.5 times the potential buoyant and other flood forces acting on an empty tank during design flood conditions.

8.5 ELEVATORS

Unless permitted otherwise in this section, all elevator components shall be located above the elevation specified in Table 8-1. Elevator components located below the elevation specified in Table 8-1 shall be capable of resisting flood-related and other loads, and constructed of flood-damage resistant materials.

The elevator pit, associated switches and hydraulic jack assembly for hydraulic elevator systems shall be permitted below the elevation specified in Table 8-1, but the electrical control panel, hydraulic pump and reservoir shall be elevated above that elevation. The hydraulic lines connecting to the hydraulic jack assembly shall be located to protect the lines from hydrostatic, hydrodynamic, debris impact, and other forces associated with the design flood.

The machine room containing the electric hoist motors and electrical control panel for traction elevator systems shall be located above the elevation specified in Table 8-1. Counterweights, roller guides, cable, pulleys, oil buffers, and electrical trip switches shall be permitted below that elevation, but shall be protected against all flood-related loads and flood-related damage, including corrosion.

Where there is the potential for an elevator cab to descend below the elevation specified in Table 8-1 during a flood event, the elevator shall be equipped with controls which will prevent the cab from descending into floodwaters.

9.0 MEANS OF EGRESS

9.1 GENERAL

Means of egress provided below the lowest floor elevations specified in Tables 2-1 and 4-1 shall be designed and constructed to resist flood-related loads and flood damage during the design flood using flood damage resistant materials.

9.2 STAIRS AND RAMPS

Stairs and ramps that provide access to elevated structures are permitted to extend into the area below the DFE. Stairs and ramps in flood hazard areas shall be designed to minimize transfer of flood-related and other loads to the structure and structure foundation. These additional loadings resulting from the obstruction presented by the stairs and ramps shall be included in the design of the structure and structure foundation.

10.0 ACCESSORY STRUCTURES

10.1 GENERAL

Accessory structures shall be designed to withstand all flood-related and other loads and load combinations, as defined in Section 1.4 of this Standard. Utilities shall be governed by Section 8.

10.2 DECKS, PORCHES, AND PATIOS

In flood hazard areas not subject to high velocity wave action, decks, porches and patios not physically attached to a structure are permitted below the elevations specified in Table 2-1, provided the decks, patios and porches are not enclosed with walls. Insect screening and open lattice panels enclosures are permitted. Decks, patios and porches shall be constructed with flood damage resistant materials in accordance with Section 6. Decks and porches less than 6 ft above eroded ground elevation shall be supported on columns or posts with a minimum dimension of 4 in.; and decks and porches 6 ft or more above eroded ground elevation shall be supported on columns or posts with a minimum dimension of 6 in. Deck and porch foundation embedment shall meet the requirements of Section 2.

In flood hazard areas subject to high velocity wave action, unattached decks, porches and patios below the elevations specified in Table 4-1 shall be designed in accordance with the requirements of Section 4.8 and constructed with flood damage resistant materials in accordance with the requirements of Section 6.

In flood hazard areas subject to high velocity wave action, decks and porches attached to a structure shall be considered to function as a continuation of the structure, and shall be:

1. designed with the lowest horizontal member of the deck or porch above the elevations specified in Table 4-1;
2. designed and constructed to withstand flood-related and other loads defined in Section 1.4, including the effects of wind, water, waves and debris impact; and

3. cantilevered from or knee-braced to the structure; cantilevered from or knee-braced to the foundation system supporting the structure; or elevated on a pile or column foundation meeting the requirements of Section 4.

Vertical supports for attached decks and porches in flood hazard areas subject to high velocity wave action shall be columns, piles or posts. Notwithstanding requirements for bearing capacity and load resistance, vertical supports shall have a minimum dimension of 6 in. and shall be spaced a minimum of 8 ft apart. Deck and porch support embedment shall be at least 6 ft below lowest eroded ground elevation, or deeper if an erosion analysis shows that 6 ft is insufficient to support the deck or porch when subjected to the load combinations of Section 1.4.2.

10.3 GARAGES

10.3.1 Garages Attached to Structures

In flood hazard areas not subject to high velocity wave action, enclosed garages attached to structures which are located below the DFE and above the lowest ground elevation shall have openings that meet the requirements of Section 2.6 and shall be constructed of flood damage resistant materials in accordance with Section 6. The garage shall be used only for parking, storage or building access. Garages shall not be located below the lowest ground elevation unless they are flood proofed in accordance with Section 7.

In flood hazard areas subject to high velocity wave action, garages attached to structures shall be constructed with breakaway walls meeting the requirements of Section 4.6.1. The garage shall be used only for parking, storage or building access. Garages located below the lowest ground elevation shall not be permitted.

10.3.2 Detached Garages

In flood hazard areas not subject to high velocity wave action, enclosed detached garages located below the DFE shall have openings that meet the requirements of Section 2.6.1 and shall be designed of flood damage resistant materials in accordance with Section 6. The garage shall be used only for parking or storage.

In flood hazard areas subject to high velocity wave action, unless properly elevated on piles or columns in accordance with Section 4, garages which are not attached to structures shall not be permitted.

10.4 CHIMNEYS AND FIREPLACES

In flood hazard areas not subject to high velocity wave action, chimneys are permitted to extend below the elevation specified in Table 8-1. Chimneys extending below the elevation specified in Table 8-1 shall be vertically supported, independent of the structure, and at a minimum on a concrete spread footing at least 12 in. thick and extending at least 6 in. beyond each side of the exterior dimension of the chimney. The thickness of the footing shall not be less than the projection. Footing depth shall meet the requirements of Section 2. The chimney shall be designed, constructed and attached to the structure to withstand all flood-related and other loads.

In flood hazard areas subject to high velocity wave action, the base of the chimney or fireplace shall not extend below the elevation specified in Table 8-1. Chimneys shall be vertically supported on piles or column foundations. Chimney foundation embedment shall be at least as deep as the rest of the structure or deeper where needed to support the chimney against flood-related and other loads. The chimney or fireplace system shall be designed to minimize transfer of flood-related and other loads or load combinations to the structure and structure foundation.

11.0 REFERENCES

All documents referenced in this standard are listed below with full titles and dates.

[1] American Society of Civil Engineers. 1995. Minimum Design Loads for Buildings and Other Structures, ASCE 7-95.
[2] American Society for Testing and Materials. 1978. Standard Test Methods for Moisture-Density Relations of Soil and Soil-Aggregate Mixtures Using 5.5-lb (2.49-kg) Rammer and 12-in. (305-mm) Drop, ASTM D698.
[3] American Society of Civil Engineers. 1995. Building Code Requirements for Masonry Structures, ACI-530/ASCE-5/TMS-402.
[4] American Concrete Institute. 1992. Building Code Requirements for Structural Concrete, ACI 318.
[5] American Society for Testing and Materials. 1992. Standard Specification for Round Timber Piles, ASTM D25.

[6] American Society for Testing and Materials. 1990. Specification for Welded and Steel Pipe Piles, ASTM A252.

[7] American Society for Testing and Materials. 1989. Specification for Zinc (Hot-Dip Galvanized) Coatings on Iron and Steel Products, ASTM A123.

[8] American Society for Testing and Materials. 1987. Specification for Zinc Coating (Hot-Dip) on Iron and Steel Hardware, ASTM A153.

[9] American Society for Testing and Materials. 1993. Specification for General Requirements for Steel Sheet, Zinc-Coated (Galvanized) by the Hot-Dip Process, ASTM A525.

[10] American Society for Testing and Materials. 1993. Specification for Structural Steel, ASTM A36/A36M.

[11] American Society for Testing and Materials. 1993. Specification for High-Strength Low-Alloy Columbium-Vanadium Steels of Structural Quality, ASTM-A572/A572M.

[12] American Society for Testing and Materials. 1993. Specification for High-Strength Low-Alloy Steel H-piles and Sheet Piling for Use in Marine Environments, ASTM-A690/A690M.

[13] National Fire Protection Association. 1994. National Electrical Code, NFPA 70. Quincy, MA.

COMMENTARY

This commentary is not part of SEI/ASCE 24-98. It is included for information purposes.

C1.0 GENERAL

C1.1 SCOPE

The requirements of this standard are developed through a rigorous consensus process of the American Society of Civil Engineers and are intended to meet or exceed the requirements of the National Flood Insurance Program (NFIP). Any conflicts or differences between this Standard and other applicable regulations should be resolved such that compliance with NFIP requirements is attained. Certain terminologies of NFIP requirements not incorporated in the Standard are provided in this commentary.

This Standard is not intended to apply to routine or minor repairs or improvements to a structure. However, it should be noted that in some instances, repairs to or improvements of an existing structure may be sufficient for the authority having jurisdiction to classify the work as a "substantial improvement." In some instances, damage to a structure may be sufficient for the authority having jurisdiction to designate that structure as "substantially damaged." In either instance, the authority having jurisdiction may require work to be in conformance with requirements for new construction. The authority having jurisdiction should be consulted to determine whether or not repairs or improvements in a flood hazard area require use of this Standard or other requirements commonly applied to new construction.

The provisions of this Standard may not apply to certain historic structures. The authority having jurisdiction and/or the State Historic Preservation Office should be consulted as to whether or not the provisions of this Standard apply.

C1.2 DEFINITIONS

Building Code: Regulations adopted by an authority having jurisdiction which set forth Standards for the design and construction of structures, for the purpose of protecting the health, safety and general welfare of the public.

Development: Any man-made change to improved or unimproved real estate, including but not limited to structures, permanent storage of materials, mining, dredging, filling, grading, paving, excavations, drilling operations and other land disturbing activities.

Federal Emergency Management Agency (FEMA): The independent federal agency that, in addition to carrying out other activities, oversees the administration of the National Flood Insurance Program.

Flood Hazard Boundary Map (FHBM): The first flood risk map prepared by FEMA for a community, which identifies flood hazard areas based on approximation of land areas in the community having a 1% or greater chance of flooding in any given year.

Federal Insurance Administration (FIA): The component of FEMA directly responsible for administering the flood insurance aspects of the National Flood Insurance Program.

Flood Insurance Rate Map (FIRM): An official map of a community, on which FEMA has delineated both the special hazard areas and the risk premium zones applicable to a community.

Flood Insurance Study (FIS): An examination, evaluation and determination of flood hazards and, if appropriate, corresponding water surface elevations, or an examination, evaluation and determination of mudslide (mudflow) and/or flood-related erosion hazards, and a report prepared by FEMA of the flood hazards studied for a community. The FIS typically includes a section on discharges of streams studied, water surface elevations for floods, and floodway data.

Floodplain Management Ordinance: An ordinance adopted by the authority having jurisdiction to regulate development in flood hazard areas.

Manufactured Housing: A structure that is transportable in one or more sections, built on a permanent chassis and constructed to Federal Mobile Home Construction and Safety Standards and rules and regulations promulgated by the U.S. Department of Housing and Urban Development.

Mitigation Directorate: The component of FEMA dedicated to the elimination or reduction of long-term risk to life and property caused by flooding and other hazards, and administers the floodplain management and hazards identification aspects of the National Flood Insurance Program.

FLOOD RESISTANT DESIGN AND CONSTRUCTION

TABLE C1-1. Hyperconcentrated Sediment Flow Behavior as a Function of Sediment Concentration

Description	Sediment Concentration by volume	Sediment Concentration by weight	Flow Characteristics
Landslide	0.65–0.80	0.83–0.91	Will not flow; failure by block sliding
	0.55–0.65	0.76–0.83	Block sliding failure with internal deformation during the slide; slow creep prior to failure
Mudflow	0.48–0.55	0.72–0.78	Flow evident; slow creep sustained mudflow; plastic deformation under its own weight; cohesive; will not spread on level surface
	0.45–0.48	0.69–0.72	Flow spreading on level surface; cohesive flow; some mixing
Mud Flood	0.40–0.45	0.65–0.69	Flow mixes easily; shows fluid properties in deformation; spread on horizontal surface but maintains an inclined fluid surface; large particle (boulder) setting; waves appear but dissipate rapidly
	0.35–0.40	0.59–0.65	Marked settling of gravels and cobbles; spreading nearly complete on horizontal surface; liquid surface with two fluid phases appears; waves travel on surface
	0.30–0.35	0.54–0.59	Separation of water on surface; waves travel easily; most sand and gravel has settled out and moves as bedload
	0.20–0.30	0.41–0.54	Distinct wave action; fluid surface; all particles resting on bed in quiescent fluid condition
Water Flood	<0.20	<0.41	Water flood with conventional suspended load and bedload

Mudslide: The definition of mudslide is consistent with the NFIP use. However, it is intended to only include classes of flow that are technically recognized as Mudflow and Mud Flood as shown below in Table C1-1. Land slides are not intended to be covered in this Standard since they are not directly flood related.

National Flood Insurance Program (NFIP): A federal program to provide relief from the impacts of flood damages in the form of federally backed flood insurance that is available to participating communities, contingent upon nonstructural flood loss reduction measures embodied in local floodplain management regulations adopted by a community. The program is administered by FEMA.

C1.3 IDENTIFICATION OF FLOOD HAZARD AREAS

The determination as to whether or not a structure lies within the Special Flood Hazard Area begins with the review of the local Flood Insurance Rate Map (FIRM) or Flood Hazard Boundary Map (FHBM) for a community. These maps are produced by FEMA as part of the NFIP's nationwide floodplain identification efforts. See [C1] for more information on using FIRMs to identify flood hazard areas. Reference [C2] provides a comprehensive overview of floodplain mapping and floodplain management issues.

Information can be obtained from technical studies produced in conjunction with FEMA mapping or from other sources, such as: the U.S. Army Corps of Engineers, the U.S. Natural Resources Conservation Service (formerly known as the U.S. Soil Conservation Service), the U.S. Geological Survey, and state or local floodplain management agencies.

In the absence of adequate information from those organizations listed above, practitioners in hydrology, hydraulics or coastal engineering can be consulted to determine the flood hazard potential for a site, with an analysis of site characteristics, with engineering calculations and with historical flood records (if available). Practitioners in hydrology and hydraulics can calculate flood elevations and floodways with HEC-1 [C3(a)] and HEC-2 [C3(b)] or HEC-RAS [C3(c)]. Practitioners in coastal engineering can calculate storm surge elevations and storm-induced erosion with a variety of existing models. It should be noted that in areas where infrequent but extreme flood events occur, the lack of sufficient historical data makes such determinations difficult and reduces the reliability of flood frequency analyses.

Flood hazard areas on the FEMA maps are referred to as "special flood hazard areas." Designations for most of these areas of special flood hazard begin with the letter, 'A' or 'V' followed by another letter or number, indicating that flood elevations and flood hazard factors have been determined. Zones beginning with an 'A' represent areas which have a 1%, or greater, probability of flooding during any year, but which are not expected to experience wave heights of 3 ft or more (high velocity wave action). Zones beginning with a 'V' represent areas which have a 1%, or greater, probability of flooding during any year, and where high velocity wave action is anticipated during the 1% flood. Zones beginning with a 'V' are also referred to as "coastal high hazard areas"; however, high velocity wave action can occur in lake or riverine floodplains under certain circumstances as well.

Some areas of special flood hazard are designated only by the letter 'A' (these are referred to as unnumbered A zones) or only by the letter 'V' (unnumbered V zones), indicating that flood elevations and flood hazard factors have not been determined. When a structure will be located in an unnumbered zone, consult with the authority having jurisdiction or with FEMA for additional guidance.

The national floodplain management standard administered by the NFIP is the 100-year flood, with a 1% chance of being equaled or exceeded in any given year. The NFIP refers to this flood as the "base flood." A "base flood elevation" (BFE)—the water surface elevation associated with the base flood (relative to the National Geodetic Vertical Datum [NGVD], the North American Vertical Datum [NAVD], or another datum)—is identified on most of the NFIP's maps.

Designers are reminded to exercise caution regarding use of the term "100-year flood." The public frequently misinterprets the term, believing that the occurrence of a 100-year flood precludes another flood of similar magnitude for 100 years. Wherever possible, the term "1% chance flood" should be used instead of 100-year flood.

All communities that participate in the NFIP (over 19,000 communities, at present) are required to adopt floodplain management regulations that comply with NFIP requirements. Designers should be aware, however, that many states and communities enforce more restrictive requirements, by adopting a higher flood elevation than the NFIP's base flood, by requiring construction to be elevated higher than the NFIP's requirements, or by imposing more restrictive siting, development or construction standards.

Frequently, a community will regulate a floodplain based on future development conditions expected in the floodplain, rather than existing floodplain conditions. The authority having jurisdiction should be consulted to determine the applicable requirements for construction in floodplain areas.

For the purposes of this Standard, the terms "design flood" and "design flood elevation" will be used to refer to the locally adopted regulatory flood and its associated water surface elevation. If the authority having jurisdiction regulates to NFIP minimum requirements, then the design flood and design flood elevation will be identical to the base flood and base flood elevation. If the authority having jurisdiction regulates to a higher flood elevation, then the design flood and design flood elevation will be greater than the base flood and base flood elevation.

It should be noted that—in coastal areas—the design flood elevation and the base flood elevation will include the effects of waves, i.e., the DFE and BFE will be established at the wave crest elevation or wave runup elevation, not the stillwater elevation. The designer is referred to Section 4 for more information on this topic.

Even if the authority having jurisdiction has adopted NFIP minimum requirements, it may still be prudent practice to design and construct a structure to a higher elevation and more conservative standard than that required. In some cases, V zones have not been mapped where high velocity wave action is known to occur (e.g., along the shorelines of the Great Lakes). In some cases mapping efforts are hindered by the fact that it is not possible to assign an accurate exceedance frequency to historical flood events, particularly in cases where an area is subject to unique hazards (alluvial fan flooding, mudflows, icejams, debris blockage of culverts and bridges, etc.). As a consequence, many hazard areas have not yet been mapped—but the hazards still exist. Design and construction in such instances should rely on historical data and take all known or likely hazards into consideration.

C1.4 LOADS IN FLOOD HAZARD AREAS

C1.4.1 General

While most loads (such as wind loads) acting on structures are typically on the order of tens of pounds per square foot, some flood-related loads (loads from high velocity waters, wave action, debris impacts and ice) may exceed typical loads by a factor of 10 to 100, or more. Extreme loads such as these make design and construction impractical and cost-prohibitive for structures that are not elevated above these hazards or situated outside the areas affected by the hazards. Furthermore, these extreme loads may be difficult to predict.

C1.5 IDENTIFICATION OF FLOODPRONE STRUCTURES

C1.5.1 General

Flood hazard determinations should be based on the most recent and accurate maps, studies and data available. Therefore, the community should always be contacted to obtain the latest information. However, the designer is warned that even the latest maps and studies may be incomplete, inaccurate or obsolete. In instances where maps, studies or other information are known by the designer to be incomplete, inaccurate or obsolete, the community may be able to assist the designer in obtaining and reviewing flood hazard data from other sources, or may be able to provide a listing of other sources which the designer can contact.

C1.5.2 Consideration for Flood Protective Works

Thousands of miles of levees and other flood protective works are in existence; have been designed, constructed and maintained so as to provide protection against design flood conditions. Many were built for agricultural, emergency or other purposes, and may only provide protection against smaller floods (e.g., 5–15 year flood frequencies). Most will not provide protection during a 100-year flood, and have not been recognized and credited as providing flood protection during the mapping of flood hazard areas by FEMA. Unfortunately, many levees and other similar works—even small and poorly maintained ones—provide a false sense of security. When these works fail or are overtopped by floods that exceed their crest elevation, rapid inundation and high velocity flows often result. Even if the works do not fail or overtop during a flood, they may cause flooding by trapping rainfall, runoff and other waters behind them. See Chapter 4 of [C4] for more information on areas behind unsafe or inadequate levees.

Only certain works that have been designed, constructed and maintained to rigorous standards have been recognized by FEMA as providing protection during a 100-year flood event, and have been considered in floodplain mapping. When recognized by FEMA, the area protected by major flood control works is removed from the special flood hazard area, and flood insurance is not required. Nevertheless, the area is identified on the flood insurance rate map as having residual flood risk from floods greater than the 100-year flood.

However, even certified works may not render the area behind as being safe for the design and construction of structures at grade. For example, the integrity of certain flood protective works may have been reduced by flooding or other events since the floodplain was mapped. More commonly, areas behind even well-designed and constructed works are known to suffer from drainage problems, especially during extreme flood conditions. See [C5] for special building standards developed by the State of Pennsylvania for areas protected by dikes or levees against 100-year flooding.

Prudent design and construction behind any flood protective works will entail the review of all available maps, studies and other information pertaining to the works and its condition, and any known drainage or ponding problems in the area.

C1.6 CLASSIFICATION OF STRUCTURES

Table 1-1 of the Standard classifies structures according to importance and potential threat to human life. The classification has been taken directly from ASCE-7 (Standard Reference [1]) and it may be different from the classifications in building codes. For the purposes of this Standard, only structures that represent a low hazard to human life and that pose little threat to other structures (during design flood conditions) should be classified as Category I. Note that while agricultural facilities are classified as Category I structures and are subject to the requirements of this Standard, certain agricultural facilities may be exempt from the elevation and floodproofing requirements in certain types of floodplains. For example, FEMA recognizes that wet floodproofing may be appropriate for some agricultural structures—farm storage structures, grain bins, corn cribs, and general purpose barns—even though these structures are not used for those purposes specifically identified in the NFIP regulations (parking, building access, storage) as being suitable for wet floodproofing. The NFIP will allow these types of agricultural structures to be wet floodproofed upon receipt of a variance from the community. FEMA Technical Bulletin 7-93 [C6] provides more information on this subject.

C2.0 BASIC REQUIREMENTS FOR FLOOD HAZARD AREAS

C2.1 SCOPE

This Section defines basic requirements for design and construction in flood hazard areas. Even if a site is mapped as an A zone on a flood hazard map, it is strongly recommended that the designer verify the lack of high velocity wave action, high velocity flows, erosion, debris effects, ice jams and other extreme flood hazards before applying the provisions of this Section. See Sections 3 and 4 of this Standard for requirements related to high risk flood hazard areas and areas susceptible to high velocity wave action. See Sections 5 through 10 for requirements related to Design, Flood-Resistant Materials, Floodproofing, Utilities, Means of Egress, and Accessory Structures.

The distinction between "subject to high velocity wave action" and "not subject to high velocity wave action" is important since wave action is capable of imparting significant additional loads on a structure. Likewise, flood-related forces that will act in high risk flood hazard areas are also capable of introducing significant additional loads.

Designing a structure to withstand wave forces or other extreme forces unnecessarily can result in significant structural, architectural and site planning modifications, and an increased cost of construction. On the other hand, failure to design for these loads when they occur may result in total or near-total loss of the structure, and its contents.

C2.2 GENERAL

It is essential that the designer understand the nature and magnitude of loads that will act on structures during design flood conditions. Furthermore, it is essential that appropriate load combinations be applied, especially for coastal areas or other areas around large bodies of water. Design wind loads may act simultaneously with design flood loads in such areas. This is usually not the case in riverine areas; however, in riverine floods of long duration, where deep flood water depths and long fetches exist, it is quite possible that wave action may be important and should be considered in design and construction.

Proper identification and/or calculation of DFEs and velocities are, of course, required before structure elevation, foundation type, and other aspects of design can be considered. Design flood depths and velocities must be specified with a reasonable degree of accuracy; otherwise, flood forces computed during the design process will be unreliable. If flood hazard maps and studies do not specify flood elevations (e.g., as in the case of unnumbered A zones) or flood discharge/velocity information, other sources of floodplain information should be consulted. Agencies listed in Section C1.3 should be contacted. In the absence of flood elevation and other data from FEMA or other agencies, registered professional engineers having the required expertise in hydrology and hydraulics should be consulted.

C2.3.1 Siting in Flood Hazard Areas

In addition to involving technical issues, siting of structures in flood hazard areas involves land use and regulatory issues. The designer should consult the local jurisdiction and other agencies with zoning, land use and regulatory authority, to ascertain local, state, and federal siting restrictions or conditions. Many jurisdictions or governing authorities may prohibit construction or other development activities in certain portions of the floodplain; many will impose specific requirements on siting, design and construction. For example, some jurisdictions may require that new construction not cause any of the following to occur: 1) alter the flood path, 2) increase the flood discharge, 3) increase the flood velocity, 4) increase the flood elevation, 5) increase the area of flood inundation, or 6) reduce available wetlands.

C2.3.2 Siting in Floodways

The floodway is generally considered to be the most hazardous portion of a riverine floodplain area: it conveys the greatest portion of the flood flow. Encroachment by construction or other development activities into the floodway can cause flood elevations to increase, and can have other adverse consequences.

Regulatory floodways are often designated by the authority having jurisdiction as areas within which construction or development activities will not be permitted unless hydrologic and hydraulic analyses demonstrate the proposed activity will result in no decrease in effective conveyance and no increase in flood levels during the design flood. Computer programs such as the U.S. Army Corps of Engineers' HEC-2 [C3(b)] or HEC-RAS [C3(c)] provide for such computations, so that carefully designed floodways will allow for sound utilization of flood hazard areas for development. The authority having jurisdiction should be contacted for a list of approved models and procedures.

Many of the flood insurance studies published by FEMA include regulatory floodways established by agreement between FEMA and the community. Since many FIRMs have floodways designated, new computations are generally not required. However, lack of a designated regulatory floodway on a FIRM or FHBM does not mean that a floodway does not exist —designers should check with the authority having jurisdiction and with FEMA before assuming flood hazards are uniform throughout the floodplain. In instances where regulatory floodways have not been designated, Section 60.3(c)(10) of the NFIP regulations requires communities to prohibit new construction, substantial improvements or other development (including fill) in Zones A1-30 and AE on the community's FIRM, unless it is demonstrated that the cumulative effect of the proposed development will not increase the water surface elevation of the base flood more than 1 ft at any point within the community. Some communities are even more restrictive—requiring demonstration that the cumulative effect of the proposed development will not increase the water surface elevation of the base flood at any point within the community. Some communities may require conveyance and/or storage compensation or other mitigation for any construction in the floodway.

Designers should check with the authority having jurisdiction to determine which development and construction activities are permitted in floodway areas, and to determine any limitations or special permit conditions associated with those activities. The designer should be advised that some states and communities severely restrict construction and development in floodways, and may prohibit residential construction entirely. For example, the State of Michigan prohibits all residential construction in floodway areas to ensure that the floodway remains uninhabited.

C2.4 ELEVATION REQUIREMENTS

A basic requirement for design and construction in flood hazard areas is that a structure should be capable of resisting flood forces and minimizing flood damages. This is usually accomplished by elevating the lowest floor to or above the design flood. NFIP regulations generally require elevation to or above the BFE. This Standard generally requires elevation at or above the DFE. DFE is greater than or equal to BFE. So this Standard is sometimes more restrictive than NFIP requirements. Note, however, that there may be significant savings in flood insurance premiums by elevating to higher than NFIP-required elevations. Note also, elevation to the DFE provides no safety factor against flooding. A safety factor can be incorporated by elevating above the DFE (i.e., by including freeboard).

The designer is cautioned that freeboard requirements included in Tables 2-1, 4-1, 6-1, 7-1 and 8-1 do not provide the same level of protection in all geographic areas. For example, 1 ft of freeboard may be equivalent to a 200-year flood in one area and a 500-year flood in another. Designers may wish to investigate the issue thoroughly before accepting this Standard's minimum freeboard requirements—additional freeboard might be desirable.

However, elevation of non-residential structures above the DFE is not always required by this Standard, for areas not subject to high velocity wave action (or other extreme flood loads—debris, ice, high velocity flows, etc.). Flood resistance can sometimes be accomplished in such areas with the lowest floor below the DFE, through floodproofing. Check with the authority having jurisdiction for any specific floodproofing requirements and restrictions. The designer is also referred to Sections 6 and 7 for guidance and requirements related to flood-resistant materials and floodproofing. The designer is cautioned, however, that locating the lowest floor below the DFE may result in increased construction and maintenance costs, and may result in substantially higher flood insurance premiums.

Design of structures situated near a flood hazard area boundary, but outside the flood hazard area, should also make use of floodproofing and other flood-resistant practices. In particular, basement excavation and construction at a level below the DFE should be accomplished using appropriate design and construction, even if the structure is outside the flood hazard area designated on the flood hazard map.

C2.4.1 Flood Hazard Areas Not Subject to High Velocity Wave Action

It should be noted that the Standard Section 2.4.1 is written such that the lowest floor of all new or substantially improved structures shall be at or above the elevations in Table 2-1, except for areas used solely for parking, building access and storage, and except for those areas floodproofed in accordance with Section 7. These requirements prohibit basements in residential construction. However, NFIP regulations permit engineered basements beneath residential structures in very limited circumstances—these "basement exceptions" are granted to a community participating in the NFIP when the community can show over-riding public interest in granting such an exception (use of engineered basements as tornado shelters, for example). As of 1997, 56 communities have received basement exceptions from the NFIP. It is not the intent of this Standard to prohibit residential basements in communities having received basement exceptions from the NFIP—in those instances, it is recommended that the community adopt this Standard with modification of Section 2.4.1 consistent with NFIP basement exception requirements.

C2.5 FOUNDATION REQUIREMENTS

C2.5.1 Geotechnical Considerations

Foundation design should be based on an accurate identification of underlying soil and rock properties. Although not required by many floodplain management ordinances or other applicable construction Standards, geotechnical investigations should be conducted at any construction site in a flood hazard area, where available geotechnical data are insufficient for design purposes. Foundation design should take all potential impacts of soil saturation, consolidation, movement, expansion and erosion into account, including the effects of long-term erosion occurring at a site.

C2.5.2 Foundation Depth

In addition to the guidance contained in Sections 3, 4 and 5 of this Standard, the designer can refer to References [C7, C8, C9], and to the body of literature dealing with bridge scour for assistance in calculating scour expected during design flood conditions (e.g., [C10]).

The theory and physical model testing for erosion and local scour determination have been developed mostly for structures in or along the banks of channels, that is, for bridges, embankments and abutments. However, the methodologies developed are applicable to different types of structures in the floodplain. Reference [C9] gives a conservative estimate of the depth of local scour effects from water flow diverted around a corner of a structure as:

$$S = d\{2.2(a/d)^{0.65}[V/(gd)^{0.50}]^{0.43}\}$$

where

S = depth of scour hole (ft);
d = depth of flow upstream of the structure (ft);
a = flow length measured normal to the overall direction of flow (ft);
V = velocity of flow approaching the structure (ft/s); and
g = acceleration of gravity 32.2 (ft/s^2).

The depth of scour should be increased by an appropriate factor, where the wall is at an angle skewed to the primary direction of flow.

The calculation for scour at posts, piers, and piles may also be calculated using the above equation, where a = diameter of post, pier, or pile (ft). For walls and thin-member foundations in flood hazard areas subject to wave action, S should be at least 5 ft. The value of S can probably be reduced in very cohesive soils, where it is likely that the scour depth will be less.

The designer is cautioned that the equation stated above may overestimate scour depths in certain instances; however, guidance for estimating scour depths accurately is difficult to obtain, given the variety of flow and sediment conditions that may exist. The designer should consult professionals experienced in hydraulic and sediment transport analyses in the geographic area of interest.

C2.5.3 Use of Fill

C2.5.3.1 Structural Fill

Structural fill should only be used in flood hazard areas not susceptible to high velocity wave action and other forces capable of eroding the fill. Structural fill should be suitable for its intended use, and should not lead to unacceptable levels of expansion, consolidation or movement; structural fill should be granular and free-draining, wherever feasible. Structural fill used for foundation support and protection should be properly designed, constructed and protected. Guidance on the protection of earth slopes to resist erosion from wave action and/or high flow velocities is available in several U.S. Army Corps of Engineers' publications, including [C7, C11, C12]. For low flood velocities (5 ft/s, or less) adjacent to structural fills, fill and slope protection is normally achieved with vegetation; protection against moderate flood velocities (5–8 ft/s) will require the use of stone or other materials. Use of structural fill for flow velocities greater than 8 ft/s may not be feasible.

C2.5.4 Use of Load Bearing Walls

Use of load bearing walls beneath the DFE will only be permitted where the flood-related forces acting on those walls are not sufficient to jeopardize the stability or integrity of the structure. In most instances, use of load bearing walls below the DFE in flood hazard areas susceptible to high velocity flow, high velocity wave action, or other destructive flood forces will likely result in significant damage to, or total destruction of, a structure during design flood conditions. Load bearing foundation walls designed and constructed to withstand high velocity flows, high velocity wave action or other destructive flood forces will not generally be affordable or cost-effective for construction less than three stories above the DFE.

In the case where load bearing foundation walls do not enclose an area below the DFE, it is recommended that the walls be oriented parallel to the direction of flood flow. This will reduce hydrodynamic loads, debris impact loads and other flood-related loads that will act during design flood conditions.

C2.5.4.1 Required Openings in Load Bearing Foundation Walls

Openings are required in some load bearing foundation walls (i.e., those enclosing a space below the DFE and not dry-floodproofed) in order to equalize hydrostatic pressures and prevent damage or collapse. The opening requirements are imposed by the NFIP.

C2.5.4.2 Openings in Breakaway Walls

Openings are also required in breakaway walls in A zones; however, openings are not required in breakaway walls governed by Section 4.6.1 (in V zones).

C2.5.5 Use of Piers, Posts, Columns or Piles

Although this Standard permits the use of piers, posts and columns to elevate structures above the DFE, many failures of these foundation elements have been observed following coastal flood events. The designer is cautioned that the use of piles may be more appropriate in coastal areas.

C2.6 ENCLOSURES BELOW THE DESIGN FLOOD ELEVATION

Enclosures below the DFE can be used for parking, building access and storage, provided the requirements of the authority having jurisdiction are satisfied. When a structure is located in areas not subject to high velocity wave action or high risk and where dry-floodproofing is not employed, openings in enclosures below the DFE are required to permit the automatic entry and exit of floodwaters. However, the designer is advised to keep use of enclosures below the DFE to a minimum. The designer is also advised that enclosures below the DFE may result in substantially higher flood insurance premiums. Insurance implications of enclosures should be investigated.

Non-breakaway enclosures below the DFE in areas subject to high velocity wave action are not permitted by the NFIP.

C2.6.1 Openings in Enclosures Below the Design Flood Elevation

Openings in enclosures below the DFE can be engineered or non-engineered, but should safely allow equalization of hydrostatic pressure outside and inside any enclosure below the DFE. Openings should be designed to meet the requirements of Section 2.6.1. The designer can refer to [C13] for an expanded discussion of opening requirements.

C2.6.1.1 Non-Engineered Openings in Enclosures Below the Design Flood Elevation

The opening requirement for non-engineered openings (1.0 square inch of opening area per square foot of enclosed area) is taken from the analysis described in Section C2.6.1, using certain assumptions regarding the flood water rate of rise, opening coefficient of discharge and factor of safety.

C2.6.1.2 Engineered Openings in Enclosures Below the Design Flood Elevation

The equations contained in Section 2.6.1.2 are modified versions of an equation contained in [C13]. That reference rearranges the terms in an equation describing flow, Q, through an opening of area, A, under a head differential, H [i.e., $Q = (c)(A)(2gH)^{0.5}$], and adds a factor of safety, to compute the opening area required to conduct flow Q:

$$A = \{Q \cdot (FS)\} \div \{38.0(c)(p)^{0.5}\}$$

where

A = net area of openings required (in.^{2});
Q = flow rate per square foot of opening area (gallons per minute);
c = coefficient of discharge;
p = hydrostatic pressure due to a 1.0 ft head differential (pounds per square inch);
FS = factor of safety, taken to be 5.0; and
38.0 = constant for the case where the rate of rise is 5.0 ft/h.

The equation yields $A = 0.83$ in.^{2} of opening per square foot of enclosed area, or approximately 1.0 in.^{2} per square foot of enclosed area (the non-engineered opening requirement), using a coefficient of discharge = 0.20 and under the assumptions described.

This Standard generalized the equation derived in [C13] to account for different rates of rise and factors of safety.

All other things equal, the relationships given in the standard yield area required for breakaway wall equal to one-fifth of the net opening area required for non-breakaway walls above.

Discharge coefficients shown in Table 2-2 were taken from or derived from standard hydraulics texts. The coefficient represents the ratio of the actual flow through an opening divided by the ideal flow, where ideal flow is given by:

$$Q = A(2gH)^{0.5}$$

and where Q = ideal flow (cfs), A = cross-sectional area of opening (sq ft), H = depth of flow at opening (ft), and g = gravitational constant (32.2 ft/s^2).

In the case of circular openings, hydraulics references tell us the discharge coefficient for a vertical circular, sharp-edged orifice will be approximately 0.60 under low head conditions, which are required here (maximum head difference across opening equal to 1.0 ft). Coefficients for the rectangular and square openings were calculated for typical opening sizes (i.e., 12 in. × 12 in. or 8 in. × 16 in., corresponding to nominal masonry unit sizes) using the ideal flow discharge relationship above and the following discharge relationship for a contracted rectangular weir:

$$Q = 3.330(L - 0.2H)H^{1.5}$$

where

Q = flow through opening in cfs;
L = horizontal length, ft; and
H = depth of flow through opening, ft.

Other shapes were assigned a coefficient of 0.30 based upon flow through a V-notch or trapezodial weir. This Standard also assumes a coefficient equal to 0.20 when the opening is partially obstructed. If the designer is certain the potential for blockage by debris is small, discharge coefficients between 0.25 and 0.60 should be used depending upon the opening shape. If the designer cannot be certain that the potential for debris blockage is small, a discharge coefficient of 0.20 should be used.

C3.0 HIGH RISK FLOOD HAZARD AREAS

C3.1 SCOPE

The nature of the hazards listed in Section 3.1 makes the identification of high risk flood hazard areas difficult, and design and construction in such areas problematic. In addition, the local authority having jurisdiction may prohibit all building in these areas. The intensity, spatial extent, duration, and probabilities associated with these hazards are difficult to predict, leading to uncertainties associated with the delineation and management of high risk flood hazard areas. Reference [C14] provides general guidance for management of high risk flood hazard areas. Unfortunately, the scenic beauty of many areas subject to high risk flood hazards attracts development interest and poses a serious challenge to floodplain managers and building officials.

Section 3 provides for construction in many high risk flood hazard areas, contingent upon the design and construction of protective works. However, the designer is cautioned that the same problems and uncertainties, identified above, that make the identification of high risk flood hazard areas difficult also make design and construction of protective works difficult. Construction of protective works, or reliance on existing works, to protect a structure in a high risk flood hazard area may be ill-advised, especially where uncertainties about flood hazards are great.

In high risk flood hazard areas, poor maintenance or improper operation of protective works and facilities, including pumping plants, can cause damage to or failure of those protective works and facilities, and will almost certainly result in destruction of the structure behind the protective works. Therefore, all protective works should include implementation of a well-conceived plan for periodic inspection, maintenance, repair and testing.

C3.2 ALLUVIAL FAN AREAS

Alluvial fan areas represent one of the most hazardous floodplain areas. Alluvial fans are geomorphic features characterized by cone- or fan-shaped deposits of boulders, gravel, sand, and fine sediments that have been eroded from upstream watersheds, and then deposited on the adjacent valley floor. Flooding that occurs on active alluvial fans is often characterized by debris and sediment-laden flows. Channel avulsion or overbank flows can result in unconfined flows on alluvial fans where flow paths are unpredictable and subject to lateral migration. In addition, these fast-moving flows present hazards associated with erosion, debris transport and deposition, and sediment transport and deposition. Alluvial fans can be found throughout the United States, but are numerous in arid and semi-arid regions.

An alluvial fan flooding hazard is indicated on a community's flood hazard map by three criteria: (a) flow path uncertainty below the hydrographic *apex*; (b) abrupt deposit and ensuing *erosion* of sediment as a stream or *debris flow* loses its competence to carry material eroded from a steeper, upstream source area; and (c) an environment where the combination of sediment availability, slope, and topography creates a hazardous condition for which elevation on fill will not reliably mitigate the risk.

Section 3.2 provides performance standards for construction of structures on alluvial fans. Section 3.2 does not present an exact technical representation of real world behavior; however, it is clear and precise regarding application. Design and construction in an alluvial fan should be predicated upon careful review of the physical features of an individual fan, best obtained from maps, photographs, historical flood data, soils data and personal observation.

Extensive damages experienced on alluvial fans generally result from floods exceeding the design parameters of flood control structures or hydraulic models. Extensive flood damages can be associated with surging whose flow hydraulics may exceed structural design specifications. Damages can be particularly extensive to structures which are located directly in the flow path.

Alluvial fan flood mitigation can take two approaches, upstream storage of flood or conveyance off the developed fan. Communities may wish to reserve a storage area immediately downfan from the apex, where existing structures occupying these areas would not be reconstructed once substantially damaged. Alluvial fan communities may also choose to restrict development in reserved flood passageways. However, unless a full fan flood control project is constructed, reserved flood passageways will not assure containment of the design flood. Some combination of the two approaches may be considered. Flood hazard avoidance on alluvial fans is strongly encouraged. In some cases, the local authority should consider restriction of development on the entire alluvial fan.

The construction of elevated structures in lower risk areas of an alluvial fan will not assure complete flood protection. Floods on alluvial fans have extremely unpredictable flow paths, so the design and construction standards only address design flood conditions including the potential for surging, channel avulsion, and debris frontal waves.

Floodproofing protection in high hazard areas near the fan apex is generally not acceptable for individual structures. Alluvial fan flood damage reduction can be achieved by constructing debris collection dams at the fan apex with maintenance plans for removal of debris after each flood. Flood conveyance channels would normally be designed to safely transport high velocity discharge from the debris dam and downstream tributaries to a disposal location such as a large river. The USACE flood protection project for the City of Los Angeles is an example of alluvial fan flood damage reduction.

Freeboard for channel or levee design to convey mud and debris flows must consider the potential for sediment deposition in the channel during the event. Freeboard in excess of 5 ft should be considered for watersheds with large boulders. Impact pressure on walls during a mudslide (ie., mudflow) event should consider the largest diameter boulder found on the alluvial fan at the elevation traveling at velocity equal to or in excess of the peak discharge velocity.

FEMA regulations regarding development on alluvial fans are contained in 44 CFR, Ch. 1, Sections 60.3(c)(7) and 60.3(c)(11), and provide alluvial fan restrictions for communities in the NFIP. Reference [C15] discusses FEMA regulations for alluvial fans.

C3.3 FLASH FLOOD AREAS

Where structures are constructed in a flash flood area, a serious risk to human life prevails. Good examples include the Big Thompson River downstream of Estes Park, Colorado, and Gatlinburg, Tennessee. Both of these locations possess steeply sloped mountain streams with limited channel capacity, and are subject to high intensity, short duration storms. Although flood warning systems have been developed to alleviate the threat to life in these areas, they cannot completely eliminate the threat due to the quickness of the flood events.

C3.3.1 Protective Works in Flash Flood Areas

Protective works in flash flood areas should provide protection against floods more severe than the 100-year flood, to significantly reduce the threat to life. Flood protection in these areas typically involves flood control reservoirs, which may be hard to justify due to their large cost, potential environmental impacts, and reservoir regulation difficulties.

C3.4 MUDSLIDE AREAS

The term "mudslide" is consistent with the NFIP regulations. It is intended to only include the classes of flow that are technically recognized as Mudflow and Mud flood as defined in Table C1-1. Landslides are not covered in this Standard since they are not flood related.

Mudflows are nonhomogeneous, nonNewtonian, viscous, transient hyperconcentrated sediment flood events whose fluid properties change dramatically as the flow progresses down slope. Viscous mudflow behavior is a function of the fluid matrix of water and fine sediments with a significant yield stress which must be exceeded to initiate motion. Mudflow areas are identified by irregular deposits, poorly sorted boulders, debris piles, natural levees and large boulders transported long distances on mild slopes. Alluvial fan sediment deposits have a significant percentage of slits and clays. Damage to structures is caused primarily from frontal wave impact, mudflow deposition and inundation, boulder impact, and lateral loading.

Mud floods are hyperconcentrated sediment flow, turbulent in nature, with essentially water flow behavior. Mud floods have no yield stress and large sediment particles will settle in a quiescent flow condition. Mud flood areas are identified by eroded channels, small levees, sorted boulders and sediments deposits with very little fine sediments (silts and clays). Mud floods will cause damage to structures primarily from frontal wave impact, inundation, boulder impact, and lateral loading.

Mudflows are hyperconcentrated flows with a sediment concentration ranging from 45 to 55% by volume. Inundation and deposits of mud, impact of mud frontal waves, flow competence (ability to transport large boulders), and high lateral loadings during mudflows can cause structures to collapse or be moved off their foundations.

Landslides may exceed 55% concentration by volume and can occur when hillsides become saturated from runoff infiltration or increased ground water levels. Slope stability analyses are required to identify potential landslide areas. In earthquake prone areas liquefaction should be considered in the hazard risk assessment. Many factors can contribute to the initiation of landslides and/or mudslides including vegetation removal, changes in soil moisture conditions, removal of the toe of slope, or increased loading. Usually, areas of potential landslides have surface evidence of this potential, or there is historical information where landslides have occurred in the vicinity.

C3.5 EROSION PRONE AREAS

Historical aerial photographs, topographic maps, nautical charts and survey data can be used to investigate erosion and estimate erosion rates. Many states already carry out such investigations, and have established construction setbacks based on long-term average annual erosion rates, especially in coastal areas. Some states have identified coastal areas subject to erosion up to and during 1% chance (100-year) storms. However, erosion prone areas could be above the floodplain. Vast areas of the floodplain and all bridge sites could be subject to erosion unless protective measures are taken.

Over the average lifetime of a wood-frame house constructed at the 30-year construction line, the likelihood of the structure meeting the Federal definition of "threatened" by long term erosion is 97%. Like flood elevations, erosion setbacks have no safety factor inherent in the design process other than larger distances. In other words, conditions only slightly worse than that estimated for the 30-year erosion construction line can cause severe damage. Coastal erosion also has a high natural variability over short distances. Therefore at least a 100-year erosion design standard is appropriate. At least thirteen states have adopted oceanfront setback lines. Eight of those use a multiplier of the long-term erosion rate. Most have common exceptions to allow construction farther seaward.

C3.6 HIGH VELOCITY FLOW AREAS

Prohibiting structures in high velocity flow areas is based on the potential for severe scour, unless costly protection measures are provided to divert high velocity flows or otherwise protect the structure [C10]. High velocity flow areas can be identified based on site location (e.g., within stream meander boundaries), historic observation of flood conditions at the site, and hydraulic analyses.

C3.8 ICEJAM AND DEBRIS AREAS

Reference [C16] provides additional guidance for design in icejam and iceflow areas.

One way to identify flood hazard areas subject to icejam and debris hazards is to consider the flood depth and velocity adjacent to structure. If upstream sources of ice and/or debris are present, and if the product of the flood depth and velocity at the site exceeds some critical value, say 4 ft^2/s, then potential icejam and debris hazards exist. The rationale for the depth-velocity criteria is based on the concept that small objects transported at velocities up to 4 ft/s, or larger objects transported at lower velocities, would not be expected to cause significant damage to a structure, or its foundation. However, the designer should consider this analysis as preliminary, and conduct a more detailed analysis at a given site, based on site conditions and structure design. The designer should also be careful to avoid a design that traps debris at the structure, thereby imposing potentially large and unaccounted for loads.

Section 5.3.3.4 of ASCE 7 (Standard Reference [1]) provides minimum impact loads that any structure governed by this Standard must resist. The designer should determine whether or not the impact loads dictated by ASCE 7, i.e., those caused by a 1,000 lb object moving at the speed of the floodwaters and acting over a 1 ft^2 area, are sufficient for a given situation.

C4.0 FLOOD HAZARD AREAS SUBJECT TO HIGH VELOCITY WAVE ACTION

C4.1 SCOPE

This Section defines the requirements for structures that are subject to damaging wave action during the design flood. Damaging wave action is generally considered to occur in those areas where conditions and water depths are sufficient to support breaking waves equal to or greater than 3 ft in height, and in areas subject to wave runup on steep slopes. The rationale for selecting the 3 ft wave criterion is described in [C17]. The necessity for considering wave runup effects in design has been established by post-flood damage investigations that found significant damage to structures not elevated above wave runup elevations. The designer is cautioned that waves less than 3 ft can cause damage to structures and should be considered when conditions are dissimilar to those in the reference.

The DFE in areas subject to high velocity wave action is at the wave crest elevation or wave runup elevation—not at the still water level.

C4.1.1 Identification of Areas Subject to High Velocity Wave Action

The wave hazard classification employed by this Standard is slightly different than that used by the National Flood Insurance Program (NFIP) in mapping flood hazard areas. NFIP floodplain mapping procedures restrict consideration of wave effects to coastal high hazard areas (i.e., V zones), although wave effects may also be important in some lake and riverine areas. This Standard intends for structures subject to damaging wave forces to be designed and constructed to resist those forces, regardless of location or the nature of the flooding. Hence, this standard is equivalent to or more restrictive than the NFIP regulations with respect to the identification of flood hazard areas subject to high velocity wave action.

Proper application of this Standard may result in designing and constructing for high velocity wave forces even though a site has been designated as an A zone by the NFIP. Designers using this Standard should verify if high velocity wave forces are a significant design issue at a site before following the requirements of Section 2 or 4. Reliance solely on the V zone and A zone designations made by the NFIP may not be sufficient, since it is possible that a structure located in an area designated by the NFIP as an A zone will be subject to damaging high velocity wave forces.

In order for a structure site to be designated "subject to high velocity wave action" the following conditions must hold:

1. water depths at the site must be sufficient to support 3-ft high waves,

2. wind or seismic forces must be capable of generating 3-ft waves, and local conditions must be such that the waves can propagate over the surface of the flood waters and reach the site.

A designer can determine whether or not water depths are sufficient to support 3-ft waves using a common approximation that states the minimum stillwater depth capable of supporting a 3-ft wave will be 1.28 times the breaking wave height (see [C7]). Using this approximation, a 3-ft wave will not occur in a stillwater depth less than approximately 3.8 ft. Any time stillwater flood depths exceed 3.8 ft, the potential for high velocity wave action exists.

However, when water depths at a site are assessed, calculations should be based on the stillwater flood elevations minus the eroded ground elevation expected at the site during the design flood event, not the pre-flood ground elevation at the site. Thus, a designer should obtain information on expected design storm and long term erosion or calculate the erosive effects of the design flood on local soils, taking into consideration the generalized erosion that may result from water and waves, and any localized erosion due to the interaction of water and waves with the structure being designed.

The presence of a 3.8 ft stillwater flood depth at a structure site is not, by itself, sufficient to classify the area as subject to high velocity wave action. In order for that designation to be accurate, the second condition stated above must also hold. In other words, the following must also occur: wind or seismic forces must be sufficient to generate waves equal to or greater than 3 ft in height; design flood water depths between the wave generation area and the site of interest must be sufficient to allow the waves to pass without being reduced in height.

In the case of wind-generated waves (the most common damaging waves), well-established procedures exist for quantifying the generation, propagation and dissipation of waves (see [C7] and [C18]). These references will allow a designer to calculate expected wave heights and to estimate wave runup at the site of interest, thus allowing the designer to determine whether or not a structure will be subject to high velocity wave action.

Due to uncertainties associated with floodplain mapping and the tendency for many areas to undergo significant shoreline changes through time (especially coastal areas near tidal inlets), it is recommended that the designer consider applying the requirements of Section 4 in areas landward of NFIP mapped V zones in the following instances:

1. where calculations show the V zone designation to be inaccurate,
2. where the shoreline shown on the flood hazard map has been modified significantly due to the effects of long-term erosion, severe storms or other factors,
3. where large bodies of water with sufficient fetch have no V zone mapped by the NFIP.

Notwithstanding the above instances, it is recommended that the designer consider application of the requirements of Section 4 over a zone landward of mapped V zones, even when the V zone mapping appears to accurately represent present-day wave hazards. This buffer zone should extend landward of the mapped V zone, reaching to the lesser of the following two conditions:

1. all land extending 500 ft landward of mapped limits of the coastal high hazard area,
2. land areas with ground elevations at or below the design flood elevation along the shoreline, and separated from a mapped V zone by a relatively narrow region with elevations above the design flood elevation (i.e., dune).

C4.2 GENERAL

Design for wave forces requires an understanding of the mechanisms by which waves can affect structures. If of sufficient height, waves can cause damage in the following ways (see Figs. C4-1 through C4-4, taken from [C19]):

1. waves may break against the side or underside of the structure, causing horizontal and vertical loads at least an order of magnitude higher than wind loads
2. waves may introduce significant drag, inertia and other forces on structural members supporting elevated structures
3. waves may "peak up" or be redirected beneath an elevated structure, and create large hydrodynamic uplift forces,
4. waves may break near the structure and run up steeply sloped ground around the structure, striking it with great force, runup is also a factor on flatter slopes,
5. waves may lead to instability by causing significant erosion of supporting fill beneath non-elevated structures, or by causing significant erosion of soil around foundation elements supporting elevated structures

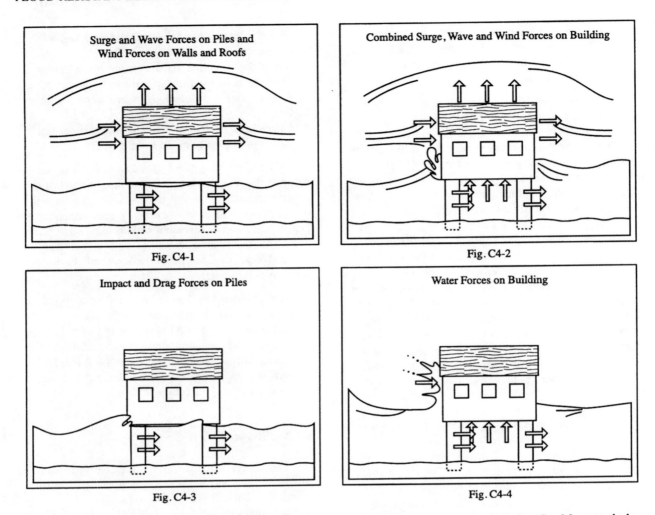

FIGURES C4-1 through C4-4. Hurricane Forces Acting on a Structure [C19]. (Reprinted with permission. Copyright 1986 Southern Building Code Congress International.)

If a designer finds that any of the above wave effects are likely to impact a structure, the design should take the anticipated wave loads and other loads during design flood conditions into consideration, especially for the design of the foundation and structural supports. It is not practical or economical to design most unelevated structures to resist wave forces. It is for this reason that stringent elevation and foundation requirements are imposed in this Standard. Guidance for design and construction of elevated structures may be found in References [C20–C22].

Specific guidance for calculation of wave loads is not presently found in ASCE 7 (Standard Reference [1]). Until such time as specific guidance is included in that document, designers should assume horizontal wave loads will be due to breaking waves, and should utilize the following formulas taken from Reference [C23]:

$$P_{max} = C_p \rho g d_s + 1.2 \rho g d_s$$

and

$$F_t = (2.41 + 1.1 C_p) \rho g d_s^2$$

where

P_{max} = maximum combined dynamic ($C_p \, \rho \, g \, d_s$) and static (1.2 $\rho \, g \, d_s$) wave pressures (psf);
F_t = total breaking wave force per unit length of structure (lb/ft);

C_p = dynamic pressure coefficient (1.60 < C_p < 3.49);
ρg = unit weight of water (lb/cu ft); and
d_s = stillwater depth (ft) at base of structure where wave breaks.

C4.3 SITING

Mangrove stands and sand dunes are natural barriers that reduce the landward transmission of waves and high velocity flows. The weakening or removal of mangrove stands or sand dunes as a result of construction or development activity can adversely impact properties by allowing waves and high velocity flows to penetrate further landward, exposing lands and structures to increased wave and hydrodynamic forces. No construction or development should take place that requires removal of sand dunes or mangrove trees, or otherwise leads to a reduction of their wave and energy dissipation characteristics.

In areas protected by sand dunes, special care must be taken during the course of any construction or development activity to prevent degradation of the structural integrity of the dune system. Manmade alteration of the dune in a V zone that can increase flood damage is prohibited by the NFIP and alteration in an A zone may also result in increased damages. Excavation of dune material should be prohibited. With the exception of dune crossovers (access ways), construction and development activities should not take place on the dune.

Extreme care should be taken in siting structures in coastal areas, in recognition of the dynamic nature of many open coast, inlet and bay/lake shorelines. Sites located near tidal inlets can experience higher than anticipated flood elevations and wave heights due to wave refraction and wave-current interactions. As a result when locating a structure near a tidal inlet, it is prudent to add an additional foot of freeboard to the elevation of the lowest supporting structural member as an additional factor of safety. Similarly, the design professional should be aware of the dynamic nature of shorelines in the vicinity of tidal inlets and attempt to identify trends or patterns in shoreline movement (using historical data) that should be considered in siting structures.

Along many shorelines, average annual long-term horizontal erosion rates have been documented at 2 to 5 ft per year, or more; in rare instances, rates exceeding 20 ft per year have been reported. Construction setbacks, based on long-term shoreline change rates, have been established by many state and local governments. Therefore, the design professional should be aware of established setback regulations and seek published long-term erosion data based on historical shoreline changes. State coastal zone management agencies typically maintain or have access to historic shoreline data and should be consulted.

C4.4 ELEVATION REQUIREMENTS

The elevation of any enclosed area used for purposes other than parking, building access and storage is set by the elevation of the lowest horizontal structural member relative to the Design Flood Elevation. The NFIP requires that the bottom of the lowest horizontal structural member be set at the base flood elevation.

This Standard has chosen to be more restrictive, taking into account building occupancy and use, as well as lowest floor framing orientation (see Table 4-1). Note that the designer must use judgment in determining the likely direction of wave approach. In most cases, waves will approach from directly offshore, or normal to the shoreline. In some cases, the likely direction of wave approach may not be easy to determine; in such cases, the designer should use the right-hand column in Table 4-1 to determine minimum elevation.

The framing orientation of the lowest floor has been determined to have a significant effect on V-zone structure performance during design flood conditions [C24]. Structures with the lowest horizontal structural supporting member satisfying NFIP elevation requirements, but oriented perpendicular to the direction of wave attack, have sustained more damage than those structures framed with the lowest supporting member parallel to the direction of wave attack. This is most likely due to the fact that in any given storm event, a certain fraction of wave heights will exceed predicted heights, damaging some structures built in compliance with NFIP elevation requirements. Therefore, this Standard imposes additional freeboard requirements for structures with lowest floor framing perpendicular to the direction of wave attack.

C4.5 FOUNDATION REQUIREMENTS

C4.5.1 General

Designers should be aware that foundations in high velocity wave action areas are often exposed to

FLOOD RESISTANT DESIGN AND CONSTRUCTION

erosion, both storm-induced and that associated with long-term recession of the shoreline. Thus, design of piling, mat and raft foundations must account for sudden and gradual erosion.

Designers should check with the local and state floodplain management officials and NFIP regarding restrictions on use of shear walls.

C4.5.3 Foundation Depth

Foundations for structures in areas subject to high velocity wave action should be designed and constructed in recognition of the dynamic nature of these areas. In addition to wave forces and flood-induced erosion, many areas are subject to long-term or cyclic erosion that should also be considered. Therefore, all new construction, substantial improvements and relocation's of existing structures in areas of flood hazard subject to high velocity wave action should be designed as if the structure exists in a highly erosion prone area. Pile foundations in high velocity wave action areas should be embedded to no shallower than -10 ft below Mean Water Level (MWL), or into a non-erodible surface. If soil conditions prevent pile tip elevations from reaching -10 ft MWL, piles should be embedded to refusal. Note, however, that refusal may not be a sufficient depth criteria if erosion can occur to the depth of refusal strata; in such situations, foundation members should be anchored to refusal strata.

Refusal should be defined by the geotechnical investigation but can be evaluated in the field by the number of blows per inch of pile penetration. For example, in the case of wood piles—solely for the purpose of determining embedment depth and not for meeting bearing requirements—refusal can be taken to mean 4 to 5 blows per inch if driven using a standard 5,000-pound hammer dropped from a minimum 3-ft lift, or using an equivalent hammer providing minimum of 15,000 ft/lbs of energy. For prestressed concrete piles, 6 to 8 blows per inch would be a reasonable value; and for steel piles, 12 to 15 blows per inch [C25]. Where the final pile depth is less than that called for by the design, a verification is needed to determine if the "as-built" foundation will still resist the anticipated loads during periods of erosion and scour.

Foundation depth shall consider both general and local scour effects. Thus, the designer should first obtain or calculate the expected grade elevation after wave- and/or current-induced general erosion at the site, and then assume at least the top remaining 5 ft of soil around a pile will be lost or rendered non-supportive due to local scour and liquefaction.

C4.5.5.1 Piles Terminating in Caps at or below Grade

The following Fig. C4-5 will aid the designer in understanding discussion of pile caps when configured for a step footing.

C4.5.5.3 Wood Piles

North Carolina has used 8×8 square piles almost exclusively since the late 1970s and the code has required them since the early 1980s. Field investigations after recent hurricanes found no apparent problems with widespread use of 8×8s where erosion around the foundation was not significant. That includes thousands of pile supported structures. Building surveys after hurricanes in other areas have found similar successes. However, as the exposed length after erosion increases and the number of stories increases, there is undoubtedly a limit where 10×10s or larger become appropriate.

The designer is cautioned that square timber finishes do not take treatment well where heartwood is exposed at the face. In addition scour is known to be greater surrounding square shapes than rounded shapes.

C4.5.6 Columns

Column-supported foundations can be anchored against pullout through the use of rock anchors or by penetration and grouting into non-erodible material. Where penetration into a non-erodible material is chosen, the hole into the non-erodible material should be bell-shaped and filled with grout or concrete such that the lower end of the mass will be wider than the upper, and so that pullout will be resisted even with minimal shrinkage of the grout. It is highly recommended that columns be supported on piles in this area. Field experience has shown that most columns were poorly constructed and have failed.

C4.5.6.3 Reinforced Concrete Columns

Concrete columns measuring less than 10 in. on a side or 12 in. in diameter have been found to be more susceptible to failure during flood events. While a smaller column may be adequate according to design calculations, numerous post-disaster structural assessments and field inspections have observed that

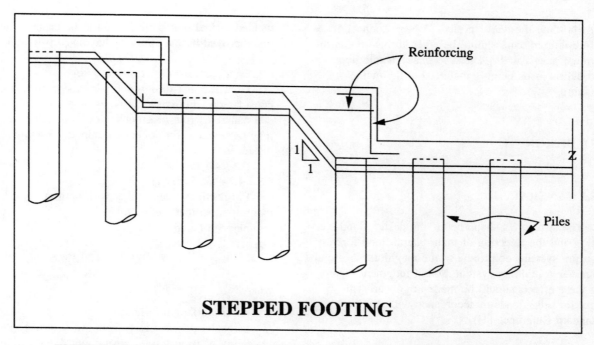

FIGURE C4-5. Typical Connections of Individual Pile Caps with Stepped Footings.

smaller size columns frequently contribute to or cause structural failures due to undetermined causes.

C4.5.8 Bracing

Cross-bracing perpendicular to the primary wave and hydrodynamic forces has been limited to rods and cables to reduce the cross-sectional area of presentment to wave forces, and because past evaluations of structure performance following major coastal storms have shown other types of cross-bracing in this direction have a high rate of failure. In designing cross-bracing, the designer should consider the likelihood of debris build-up against the bracing which will increase the surface area for wave forces to act upon and increase the loads to be resisted.

In general, it is recommended the designer strive to establish a stable design free of bracing, and to only use bracing to add rigidity to the design for the comfort of the occupants. Past experience has shown cross-bracing often fails during a storm event and does not provide the degree of support anticipated. Knee bracing has been shown to be more reliable and is recommended as the first type of bracing to be considered.

When using steel rods for cross-bracing in a highly corrosive or marine environment, the rods should be hot-dipped galvanized or fabricated from non-corrosive material.

C4.6 ENCLOSED AREAS BELOW DESIGN FLOOD ELEVATION

The NFIP requires that the use of enclosed areas below the design flood elevation be restricted to parking, access and storage; lower areas must not be finished or used for any other purpose. No mechanical, electrical or plumbing equipment, exclusive of risers, should be installed below the design flood elevation.

Though the NFIP places no restrictions on the size of enclosed areas below the design flood elevation, the designer is cautioned that the size of the enclosed area can affect flood insurance premiums. The designer may wish to contact an insurance agent to obtain information regarding enclosures and additional insurance premium charges.

C4.6.1 Breakaway Walls

The area below the design flood elevation may only be enclosed with breakaway elements. The breakaway feature can be accomplished either through material failure or through failure of connections holding the walls in place. Where connection failure is used, care should be taken to insure that individual panels will not lead to unacceptable flood and debris loads being transferred to the main structure.

C5.0 DESIGN

C5.1 GENERAL

The structure's shape below the design flood elevation and the selection of main lateral flood-force resisting systems contribute to the magnitude of flood forces on a building system. Sound judgment regarding these effects should be made in concert with those for other loads in accordance with ASCE-7 (Standard Reference [1]).

C5.2 VERTICAL FOUNDATION SYSTEMS

In mid-rise structures greater than about four stories, wind may be a controlling factor in the design of the structural frame, but flooding may play a significant role in the design of the lower walls of the structure. In structures of three or four stories, the effects of winds and potential wave action will both be a major consideration. For taller structures, forces from wind become increasingly greater and more important than those generated by flooding, and usually determine the governing design parameters for the structure. Foundation and framing systems must be able to resist the vertical forces from dead and live loads as well as lateral forces from wind and flooding. While shear walls may be appropriate for use in mid-rise structures or higher, they must be properly engineered to resist all load combinations.

C5.2.4 Piles

C5.2.4.4 Concrete-Filled Steel Pipe Piles and Shells

A minimum material thickness for steel pipe piles, shells, and casings in excess of that required by service conditions may be required to resist high collapsing pressures at the lower portion of a pile. A reduced size of coarse aggregate is often necessary for filling pile shells, and should be specified accordingly. Pile shell seams should be welded unless non-welded locked seams are proven to be capable, and are approved by the authority having jurisdiction.

C5.2.4.6 Cast-in-Place Concrete Piles

Concrete mix design and pile design should be considered carefully. Some types of piles require special concrete grout or grout mixes, and special mixes may be required for concrete pumped substantial distances.

C5.2.4.7 Pile Capacity

For sections freestanding with bottom embedment, the member can be treated as a column having an unbraced length:

$$L = H + d/12$$

where

L = unbraced length in feet;
H = height of member in ft from top of grade, plus depth of local scour, erosion, and liquefaction, to the underside of cap or other floor structure above;
d = depth of embedment, in inches, defined as the distance from top of grade minus depth of local scour and liquefaction to the point of fixity, computed conservatively as:

$$d = 1.8(EI/n_h)^{1/5}$$

where

E = modulus of elasticity of the pile (lbs/in^2);
I = minimum moment of inertia of pile (in^4); and
n_h = coefficient of horizontal subgrade reactions (lbs/in^3).

An appropriate effective length factor K, a function of end restraint, shall be applied to the unbraced length L to obtain an effective length $K(L)$.

In some situations the use of the coefficient of horizontal subgrade reaction may be inappropriate due to long-term creep effects. Other considerations for horizontal subgrade reactions should be made per a geotechnical investigation.

Minimum lateral resistance of an individual pile should be at least 5% of the design axial load due to inevitable eccentricities, lateral forces, and being out of plumb.

C5.2.4.8 Capacity of the Supporting Soils

The distribution of subgrade reaction on a pile depends on both the ratio of the pressure on the pile to a corresponding displacement, and the flexural rigidity of the pile. The computations required to estimate these are cumbersome and subject to error. The error is increased with smaller design lateral loads and larger assumed soil diameters (diameter of subsoils centered on the pile that can be assumed available to react against a pile). In lieu of more exact analyses the soil diameter is limited to three times the diameter of the pile.

C5.2.4.19 Pile Splicing

Splicing of piles is difficult and, if not done properly, can compromise the structural integrity of a foundation. Given the uncertainties related to scour of erodible soils in areas subject to high-velocity wave action and in high risk flood hazard areas, timber pile splicing should be avoided. In instances where timber piles must be spliced, the splice should be below the depth of erosion and scour predicted by an erosion analysis. In coastal areas, pile splices should be below -10 ft MWL.

C5.2.5 Posts, Piers, and Columns

Due to frequent difficulties in quality control of site-constructed piers, posts and columns in residential development, piles are recommended in lieu of posts, piers and columns. Posts, piers and columns may not be desirable in erosion- or wave-prone structures.

C5.3 FOOTINGS

Footings may not be desirable in erosion- or wave-prone structures.

C5.5 GRADE BEAMS

In high risk flood hazard areas, or in areas subject to high velocity wave action, the use of grade beams may increase the forces transmitted from the foundation to the structure, and may increase scour around the foundation. However, they are often used in engineered structures to strengthen piling or other vertical members during the most severe flood conditions. The increased scour and increased loads should be mitigated by proper design of the pile system, grade beam, and structural frame.

C6.0 MATERIALS

C6.1 GENERAL

Several sources of information are available to assist in the selection of materials for flood resistant design and construction. These include References [C26–C28].

C6.2.1 Metal Connectors and Fasteners

Exposed metal connectors and fasteners should be designed to a higher safety or load factor in coastal areas susceptible to salt spray to account for loss of material due to corrosion. Even when protected by a galvanized coating, corrosion has been shown to reduce their strength and effectiveness in just a few years. At a minimum, building owners should undertake periodic inspection and maintenance, including painting or replacement, of connectors and fasteners. The designer should strongly consider specifying stainless steel connectors and fasteners, or nonmetallic connectors and fasteners that satisfy the requirements of this Standard. Also, refer to FIA-TB-8 *Corrosion Protection for Metal Connectors in Coastal Areas* [C29].

Applications in highly corrosive environments such as near bodies of salt water with regular, breaking waves that are subject to salt laden air require additional corrosion protection and depending on the severity of the application should be either:

1. 304 or 316 alloy stainless steel, in accordance with ASTM A-276 *Specifications for Stainless and Heat-Resisting Steel Bars and Shapes* [C30], or
2. Have thicker hot-dip galvanizing as provided in FIA-TB-8 [C29].

C6.2.2 Structural Steel

Exposed structural steel shapes, beams, pipes, channels, and angles should be avoided in flood hazard areas, wherever possible, especially in coastal areas. Even when protected by a galvanized coating or other means, corrosion has been observed to occur rapidly.

C6.2.3 Concrete

Designers may wish to refer to ACI 350 *Environmental Engineering Concrete Structures* [C31] when designing concrete structures in flood hazard areas. While the requirements may be conservative for many applications, they may be useful for certain structures.

C6.2.5 Wood and Timber

Preservative treatment of wood should conform to AWPA Standards [C32–C45].

C6.2.6 Finishes

The requirement to withstand inundation for a minimum of 72 h is taken from FIA Technical Bulletin 2-93 [C27].

C7.0 DRY AND WET FLOODPROOFING

C7.1 SCOPE

Floodproofing measures have been used to protect many new and existing structures located in flood hazard areas. Such measures may completely protect a structure from the design flood, or offer protection against smaller or larger floods.

Planning and design of floodproofing measures should consider the implications, including flood insurance premium charges, of floodproofing to an elevation higher than the NFIP required elevation.

Relocation and elevation measures are the most likely to floodproof structures; however, other floodproofing measures, when properly designed, can also provide flood protection. Floodproofing measures built into a structure and requiring no human intervention during the flood event are the most reliable and preferred method of floodproofing for all structures located in the flood hazard area. It is a common practice to floodproof structures to a height of 2 to 3 ft above grade, although higher levels of floodproofing can be engineered.

In Special Flood Hazard Areas, the NFIP regulations require individual property owners to elevate a structure to or above the Base Flood Elevation except in A Zones where non-residential structures may be floodproofed. In V Zones, structures must be elevated on open foundation systems such as piles, columns, piers, or posts. In A Zones all the previously mentioned elevation techniques, plus solid perimeter foundation walls and the use of structural fill is permitted under the NFIP regulations. Additionally the NFIP requires, in mapped floodways, all development, including elevating a structure, must not result in a rise in the BFE.

Design of levees and floodwalls intended to provide flood protection to an individual or group of structures are outside the scope of this Standard. There are numerous engineering, non-engineering, and regulatory issues involved in the design and construction of levees and floodwalls. Therefore, caution should be exercised when considering employing a levee or floodwall as a method of providing flood protection. Local floodplain management officials should be consulted.

There are many documents which provide detailed information on planning, design, and construction of these floodproofing measures. Among the more notable are the USACE's *Flood Proofing Regulations* [C26] and the *Colorado Floodproofing Manual* [C46], and two FEMA publications: *Floodproofing Non-Residential Structures* [C47] and *Engineering Principles and Practices for Retrofitting Flood Prone Residential Buildings* [C48]. These documents provide information for engineers, architects, city officials, and building owners regarding all aspects of floodproofing existing structures in flood hazard areas.

NFIP regulations regarding floodproofing, contained in 44 CFR Ch. 1 subparagraphs 60.3C(3), (4), and (8), should also be consulted. They provide floodproofing criteria required under the NFIP. These regulations are discussed in FEMA Technical Bulletin 3-93, *Non-Residential Floodproofing—Requirement and Certification* [C49].

C7.2 DRY FLOODPROOFING

Whenever dry floodproofing is proposed for the lowest story of a new structure, whether it be above grade, below grade, or a combination of the two, assurance must be provided that reliable flood protection will be achieved, and that the structure will be substantially impermeable to the passage of floodwater, against all floods up to and including the design flood. This requires strict adherence to materials and construction requirements for dry floodproofing.

C7.2.1 Dry Floodproofing Restrictions

Dry floodproofing in residential structures is not permitted because it frequently requires human action such as installing flood shields, maintaining the protective features, having an operational plan, and being able to take actions within a reasonable warning time. The possible failure of the home owner to take such an action either because of absence from home, lack of maintenance, change in ownership of the home, is regarded as an unacceptable risk.

The 5.0 ft/s velocity restriction for floodproofing is not a FEMA requirement but is used in the U. S.

Army Corps of Engineers regulations [C26] in design of structures exposed to water loads from stagnant or flowing waters. Although effective dry floodproofing can be designed for higher velocities, this is a reasonable existing limit that addresses safety of dry floodproofed structures during a flood.

C7.3 WET FLOODPROOFING

Wet floodproofing techniques are used to reduce flood damages when an enclosed area of a structure is designed to allow entry and exit of floodwaters. However, all construction materials below the DFE should be flood-resistant and have appropriate structural strength to resist flood forces. FEMA Technical Bulletin 7-93, *Wet Floodproofing Requirements for Structures Located in Special Flood Hazard Areas* [C6] provides information on NFIP regulations regarding floodproofing. The areas for which wet floodproofing is permitted includes enclosed areas used solely for building access, parking, and storage as well as certain types of agriculture structures. Some examples of enclosed areas that may be wet floodproofed include garages and areas beneath elevated buildings.

C7.4 ACTIVE FLOODPROOFING

In some instances, active floodproofing provides an economic and appropriate means of floodproofing a structure. However, human intervention is required and strict guidelines for implementing active floodproofing must be followed (persons responsible for installing or implementing active floodproofing must be familiar with the procedures and equipment; sufficient warning time must be given to insure the measures are put into place, etc.). Guidance on flood warnings is available in a document entitled *Guidelines on Community Local Flood Warning and Response Systems* [C50].

C8.0 UTILITIES

C8.1 GENERAL

Standard installation practices for electric, plumbing, mechanical, HVAC and other systems often render them exposed and vulnerable to flood damage. Flood damage to utilities and attendant equipment may result in effluent discharge from sewer lines, contamination of potable water, fire hazards from damage to gas or electric lines, and other preventable losses. Further, disruption of utility services can cause a structure to be uninhabitable after a flood, even if the structure is otherwise intact. The provisions of this section apply to all utilities unless specifically exempted by the authority having jurisdiction as expendable and unlikely to endanger other components or structures in the event of damage to the utility. Where utilities and attendant mechanical equipment are permitted below the elevation in Table 8-1, a good practice is to place utilities on the most sheltered side of the foundation members.

Protective floodproofed enclosures intended to protect mechanical, electrical, and utility systems larger than 2 ft in size (as measured in any dimension) and exposed to flood forces begin to compromise the unobstructed flow requirements below elevated structures in V-zones, and may increase transmission of additional loading onto the foundation system. This will pose problems in high risk flood hazard areas. Reference [C20] provides guidance on supporting exterior platforms for utilities and attendant mechanical equipment on piles or columns.

C8.2 ELECTRICAL

Overhead electrical service may be unsightly, but moving the power lines out of the reach of flood and water damage may outweigh the aesthetic advantage of buried lines. However, overhead lines should be designed with wind forces and potential debris effects in mind. Care must be taken in the fastening overhead incoming power service to the structure, since cases have been observed where the downed wires have removed sheathing allowing rain entry into structures (see [C20]). The local utility company should be contacted for design and installation requirements related to wind. The National Electrical Code (Standard Reference [13]) should be consulted regarding burial requirements for electrical lines.

C8.5 ELEVATORS

Reference [C51] provides guidance from the NFIP related to elevator installations in flood hazard areas. Much of the information contained in Section C8.5 was taken from that document.

There are two main types of elevators, hydraulic and traction. The hydraulic elevator consists of a cab attached to the top of a hydraulic jack assembly which normally extends below the lowest floor and is operated by a hydraulic pump and reservoir. The traction elevator, the more common elevator found in structures over three or four stories, consists of a cable that is connected to the top of the cab and is operated by an electric motor located above the elevator shaft.

Hydraulic elevators will often require part of the assembly to be located below the DFE since the jack assembly is located below the lowest floor. The hydraulic pump and reservoirs, however, can be easily located up to two floors above the jack, and thus usually above the DFE. Although the casing around the jack is usually resistant to small amounts of water seepage, total inundation by floodwaters, will usually result in contamination of the hydraulic oil and possible damage to the cylinders and seals of the jack. Salt water is particularly damaging due to its corrosive nature.

Potential flood damage to elevators can be dramatically reduced when using traction elevators since the majority of the equipment is normally located above the elevator shaft and therefore not in contact with flood waters. When some equipment cannot be located above the DFE, such as oil buffers, compensation cable and pulleys, and roller guides which usually are located at the bottom of the shaft, flood resistant materials should be used where possible.

The elevator cab, common to both elevator types, typically automatically descends to the lowest floor upon loss of electrical power in accordance with fire safety code requirements. To minimize loss of life, injuries and damages, a system of interlocking controls with one or more float switches in the elevator shaft which prevents the cab from descending into floodwaters should be installed. In addition, the majority of the electrical equipment for both elevator types, such as electrical junction boxes and circuit and control panels, can be elevated above the DFE.

C9.0 MEANS OF EGRESS

C9.1 STAIRS AND RAMPS

Stairs and ramps located in flood hazard areas subject to wave action or high velocity flow should be designed to transfer minimal loads to the foundation and structure supported above. Orientation, open rails, open or designed breakaway risers, and retractable stairs are just a few methods of reducing load impacts.

When stairs are oriented parallel to the flow of the floodwaters, risers should be open or designed to breakaway to allow for the passage of floodwaters through the stairway. Stairs and ramps located below the DFE should be designed with open guardrails to allow for the passage of flood waters.

An alternative to traditional in-place stair design, are retractable stairs which can be "raised" above the DFE. The top of the stairway which is located at or above the DFE is hinged; the bottom of the stairway is attached to a pulley-type system so that the bottom of the stairway may be raised to or above the DFE, thus removing the stairway from the flow path of the floodwaters. These stairs should be designed in compliance with local building codes.

C10.0 ACCESSORY STRUCTURES

C10.1 DECKS, PORCHES, AND PATIOS

In areas of flood hazard subject to high velocity wave action, results of post-storm evaluations and surveys have indicated that elevated decks and porches are often founded on insufficiently embedded columns or piles. The decks or porches become undermined, causing them to crash into the structure or collapse and become susceptible to wave attack which breaks them apart causing an increase in debris damage to the structure. Therefore, since areas of flood hazard subject to wave action or high velocity flows are often highly susceptible to erosion, adequate embedment depth is integral to the survival of the deck, porch and patio.

Cantilevered and knee-braced decks and porches are recommended for areas subject to erosion since the structural elements (piles, columns, etc.) rely on support from the primary foundation system which is required to be set much deeper making damage due to erosion less likely. Where the deck is to rely on its own supporting foundation, it is recommended the foundation meet the design requirements of the primary structure.

C10.3 GARAGES

C10.3.1 Garages Attached to Structures
Enclosed garages attached to structures located below the DFE and above the lowest ground elevation in areas not subject to high velocity wave action

require openings to allow balancing of hydrostatic forces. Entry and exit of floodwaters is not ensured in gaps between the door segments and the garage door and the garage door jamb. In addition, opening requirements are not satisfied when human judgment and action are needed to open the garage door or other doors prior to flooding.

Enclosed garages attached to structures, and located below the lowest ground elevation in areas are considered a basement. Although not permitted below residential structures, NFIP regulations permit garages to be located below the lowest ground elevation for non-residential and mixed-use structures, provided the structures are not subject to high velocity flow and provided the garages are dry floodproofed. In most designs, the loadings on the above-grade portion of the structure are transferred to the structural elements of the below-grade garage. Thus, any structural failure of the garage may result in failure of the entire structure. One of the critical elements in the floodproofing design of the garage below ground elevation is the point where the garage entrance ramp meets the street grade. The optimum design would be to design the entrance at or above the DFE. However since the garage entry must often meet street grade elevations that are below the DFE, the garage should be designed with a high-strength flood shield that can withstand the high hydrostatic pressures to prevent floodwaters from entering the dry floodproofed garage. In addition, a sufficient number of emergency exits should be available so that anyone in the garage will not be trapped by rising floodwaters.

C10.4 CHIMNEYS AND FIREPLACES

Chimneys and fireplaces are typically either masonry or factory-built. Similar to decks and porches in areas subject to high velocity wave action and erosion, chimneys are often supported on inadequately embedded foundations that are undermined, thus causing the chimney to crack or separate from the structure. Therefore, adequately embedded foundations are essential to the reduction in flood damage to chimneys. It is recommended that chimneys be elevated above the design flood level, even in areas not subject to wave action.

C11.0 REFERENCES

[C1] Federal Emergency Management Agency. 1988. Guide to Flood Insurance Rate Maps, FIA-14. Washington, D.C.

[C2] Federal Emergency Management Agency. 1992. Floodplain Management in the United States: an Assessment Report, FIA-18, Washington, D.C.

[C3(a)] U.S. Army Corps of Engineers, Hydrologic Engineering Center, Computer Program, HEC-1, Flood Hydrograph Package, Version 4.0, September 1990.

[C3(b)] U.S. Army Corps of Engineers, Hydrologic Engineering Center, Computer Program, HEC-2, Water Surface Profiles, Version 4.6, May 1991.

[C3(c)] U.S. Army Corps of Engineers, Hydrologic Engineering Center, Computer Program, HEC-RAS, River Analysis System, Version 2.0, April 1997.

[C4] Federal Emergency Management Agency. 1987. Reducing Losses in High Risk Flood Hazard Areas: A Guidebook for Local Officials, FEMA 116. Washington, D.C.

[C5] Pennsylvania Department of Community Affairs. 1981. Handbook of Flood Resistant Construction Specifications, Suggested for Use in Areas Protected by Dikes and Levees. Harrisburg, PA.

[C6] Federal Emergency Management Agency. 1993. Wet Floodproofing Requirements for Buildings Located in Special Flood Hazard Areas, National Flood Insurance Program Technical Bulletin 7-93, Washington, D.C.

[C7] U.S. Army Corps of Engineers. 1984. Shore Protection Manual, Washington, D.C.

[C8] Fowler, Jimmy E. 1993. Coastal Scour Problems and Prediction of Maximum Scour, Technical Report CERC-93-8, U.S. Army Corps of Engineers, Washington, D.C.

[C9] Simons, D.B. and F. Senturk. 1977. Sediment Transport Technology. Water Resources Publications, Fort Collins, CO.

[C10] Federal Highway Administration. 1988. Interim Procedures for Evaluating Scour at Bridges. Office of Engineering, Bridge Division.

[C11] U.S. Army Corps of Engineers. 1994. Earth and Rock-Fill Dams, Engineer Manual EM 1110-2-2300, Washington, D.C.

[C12] U.S. Army Corps of Engineers. 1978. Design and Construction of Levees, Engineer Manual EM 1110-2-1913, Washington, D.C.

[C13] Federal Emergency Management Agency. 1993. Openings in Foundation Walls for Buildings Located in Special Flood Hazard

Areas, National Flood Insurance Program Technical Bulletin 1-93, Washington, D.C.

[C14] Federal Emergency Management Agency. 1987. Reducing Losses in High Risk Flood Hazard Areas: A Guidebook for Local Officials, FEMA-116, Washington, D.C.

[C15] Federal Emergency Management Agency. 1989. Alluvial Fans: Hazards and Management, FEMA-165, Washington, D.C.

[C16] U.S. Army Corps of Engineers. 1982. Engineering and Design: Ice Engineering, Engineer Manual EM 1110-2-1612, Washington, D.C.

[C17] U.S. Army Corps of Engineers, Galveston District. 1975. Guidelines for Identifying Coastal High Hazard Zones.

[C18] National Academy of Sciences. 1977. Methodology for Calculating Wave Action Effects Associated with Storm Surges.

[C19] Southern Building Code Congress International. 1986. Hurricane Resistant Construction Manual.

[C20] Federal Emergency Management Agency. 1986. Coastal Construction Manual, FEMA 55, Washington, D.C.

[C21] Federal Emergency Management Agency. 1984. Elevated Residential Structures, FEMA 54, Washington, D.C.

[C22] Federal Emergency Management Agency. 1985. Manufactured Home Installation in Flood Hazard Areas, FEMA 85, Washington, D.C.

[C23] Walton, T. L., Jr., J. P. Ahrens, C. L. Truitt and R.G. Dean. 1989. *Criteria for Evaluating Coastal Flood Protection Structures,* Technical Report CERC 89-15. U.S. Army Corps of Engineers, Waterways Experiment Station, Vicksburg, MS.

[C24] Coastal Planning & Engineering/URS Consultants, Inc. 1992. Coastal Structure Damage Methodology: Final Report. submitted to the U.S. Army Corps of Engineers, Philadelphia District.

[C25] Bowles, J.E. Foundation Analysis and Design. 1977. McGraw Hill, Inc., New York.

[C26] U.S. Army Corps of Engineers. 1995. Flood Proofing Regulations, EP 1165-2-314. Washington, D.C.

[C27] Federal Emergency Management Agency, 1993. Flood-Resistant Materials Requirements for Buildings Located in Special Flood Hazard Areas, National Flood Insurance Program Technical Bulletin 2-93. Washington, D.C.

[C28] Texas Coastal and Marine Council. 1981. Model Minimum Hurricane Resistant Building Standards for the Texas Gulf Coast, Austin, TX.

[C29] Federal Emergency Management Agency. 1993. Corrosion Protection for Metal Connectors in Coastal Areas, National Flood Insurance Program Technical Bulletin 8-93, Washington, D.C.

[C30] American Society for Testing and Materials. 1992. Specification for Stainless and Heat-Resisting Steel Bars and Shapes, ASTM A276.

[C31] American Concrete Institute. 1989. Environmental Engineering Concrete Structures, ACI-350.

[C32] American Wood Preservers' Association. 1988. All Timber Products—Preservative Treatment by Pressure Processes, AWPA C1-88.

[C33] American Wood Preservers' Association. 1989. Lumber, Timber, Bridge Ties and Mine Ties—Preservative Treatment by Pressure Processes, AWPA C2-89.

[C34] American Wood Preservers' Association. 1985. Plywood—Preservative Treatment by Pressure Processes, AWPA C9-85.

[C35] American Wood Preservers' Association. 1989. Piles—Preservative Treatment by Pressure Processes, AWPA C3-89.

[C36] American Wood Preservers' Association. 1989. Poles—Preservative Treatment by Pressure Processes, AWPA C4-89.

[C37] American Wood Preservers' Association. 1988. Dual Treatment, MP1-88.

[C38] American Wood Preservers' Association. 1988. Creosote, MP2-88.

[C39] American Wood Preservers' Association. 1988. Waterborne Salts, MP4-88.

[C40] American Wood Preservers' Association. 1988. Creosote, AWPA P1-88.

[C41] American Wood Preservers' Association. 1988. Creosote-Coal Tar Solutions, AWPA P2-88.

[C42] American Wood Preservers' Association. 1988. Waterborne Preservatives, AWPA P5-88.

[C43] American Wood Preservers' Association. 1988. Pentachlorophenol, AWPA P8-88.

[C44] American Wood Preservers' Association. 1987. Pentachlorophenol, AWPA P9-87.

[C45] American Wood Preservers' Association. 1987. Foundation Piles, AWPA FND-88.

[C46] Colorado Water Conservation Board. 1983. Colorado Floodproofing Manual.

[C47] Federal Emergency Management Agency. 1986. Floodproofing Non-Residential Structures, FEMA 102, Washington, D.C.

[C48] Federal Emergency Management Agency. 1995. Engineering Principles and Practices for Retrofitting Flood Prone Residential Buildings, FEMA 259. Washington, D.C.

[C49] Federal Emergency Management Agency. 1993. Non-Residential Floodproofing—Requirements and Certification for Buildings Located in Special Flood Hazard Areas, National Flood Insurance Program Technical Bulletin 3-93. Washington, D.C.

[C50] Federal Interagency Advisory Committee on Water Data, Hydrology Subcommittee. 1985. Guidelines on Community Local Flood Warning and Response Systems. Reston, Virginia.

[C51] Federal Emergency Management Agency. 1993. Elevator Installation for Buildings Located in Special Flood Hazard Areas, National Flood Insurance Program Technical Bulletin 4-93. Washington, D.C.

INDEX

A zone 1, 31
Accessory structures; design and construction guidelines for 26–27, 50–51
Active floodproofing 23, 49
Adjacent grace 1
Air conditioning; design and construction guidelines for 25
Alluvial fan 1
Alluvial fan areas 11, 38
Alluvial fan flooding 1, 38
Anchorage 20
Apex 1, 38
Attendant utilities and equipment 1, 24–25, 49–50
Authority having jurisdiction 1

Base flood 1, 31
Base flood elevation (BFE) 1, 31, 34
Basement 1
Bedrock 1
Bracing 17, 18, 45
Braided channel 1
Breakaway walls; defined 1; design guidelines for 17, 45–46; openings in 9, 36
Building code 29
Bulkheads 1, 17

Cast-in-place concrete piles 18, 46
Channel 1
Channel stabilization 1
Check valve 1
Chimneys; design and construction guidelines for 27, 51
Circuit breakers 24
Coastal High Hazard Area (CHHA) 1
Columns; design guidelines for 9–10, 20, 36, 47; in high velocity wave action areas 13, 16, 44–45
Combination of loads 6
Community 1
Concrete; materials requirements for 22, 47
Concrete pads; design guidelines for 17
Concrete piles 18, 20, 46
Concrete walls; design requirements for 18
Concrete-filled steel pipe piles 18, 46
Connectors; materials requirements for 21, 47
Conveyance 1
Corrosive environment; for structural steel 21
Cross-bracing 17, 45

Debris flow 1, 38
Debris impact loads 1–2
Decks; design guidelines for 17, 26–27, 50

Design flood 2, 31
Design flood elevation (DFE) 2, 31, 34
Development 29
Disconnect switches 24
Dry floodproofing 2, 22–23, 48–49
Ductwork; design and construction guidelines for 25

Electric meters 24
Electrical cable 24, 49
Electrical conduit 24, 49
Electrical utilities; design and construction guidelines for 24–25, 49
Elevation; base flood elevation 1, 31; design flood elevation (DFE) 2; eroded ground elevation 2; stillwater elevation 5; water surface elevation 5; wave crest elevation 5, 31; wave runup elevation 5, 13, 31
Elevation requirements; in flood hazard areas 8, 34–35; in high velocity wave action area 8, 13, 43; NFIP standard 43
Elevators; design and construction guidelines for 26, 49–50
Enclosures 2
Enclosures below the design flood elevation 10–11, 17, 36–37, 45–46
Engineered openings; in enclosures below design flood elevation 10, 36–37
Eroded ground elevation 2
Erodible soil 2
Erosion 2, 38, 41
Erosion analysis 2–3
Erosion control structures 17
Erosion prone areas; design and construction guidelines for 12, 39
Existing structure 3

Fasteners; materials requirements for 21, 47
Federal Emergency Management Agency (FEMA) 29
Federal Insurance Administration (FIA) 29
Fetch 3
Fill 3, 9, 15, 35–36
Finishes; materials requirements for 22, 48
Fireplaces; design and construction guidelines for 27, 51
Flap gate 3
Flash flood areas 11, 38–39
Flash flow 3
Flood 3, 30
Flood control structure 3
Flood fringe 3

Flood hazard areas 33; alluvial fan areas 11, 38; defined 3; elevation requirements for 8, 13, 34–35, 43; erosion prone areas 12, 39; flash flood areas 11, 38–39; floodproofing 2, 3, 5, 22–23, 48–49; foundation requirements for 8–10, 13, 15–17, 35–37, 43–45; high risk 11–12, 37–40; high velocity flow areas 3, 12, 40; high velocity wave action areas 8, 12–13, 40–45; icejam and debris areas 4, 12, 40; identification of 6, 30–31; loads in 6, 10, 31–32; materials requirements for 21–22, 47–48; mudslide areas 11–12, 39; siting in 8, 13, 33, 43; special flood hazard areas 5, 30–31, 48

Flood Hazard Boundary Map (FHBM) 29, 30
Flood hazard map 3
Flood hazard study 3
Flood Insurance Rate Map (FIRM) 30
Flood Insurance Study (FIS) 29
Flood protective works 3, 6–7, 32
Flood routing 3
Flood-damage-resistant material 3
Flood-related erosion 3
Flooding 3
Floodplain 3
Floodplain management ordinance 29
Floodprone structures; identification 6–7, 32
Floodproofing 3, 22, 48–49; active 23, 49; alluvial fan flood damage 38; dry 2, 22–23, 48–49; wet 5, 23, 49
Floodwalls 3, 48
Floodways; defined 3; siting in 8, 33–34
Footings; defined 3; design guidelines for 20, 47; spread footings 13
Footprint 3
Foundation requirements 8–10, 35–37, 43–45; bracing 17, 45; columns 9–10, 13, 16, 36, 44–45, 47; fill 9, 13, 35–36; foundation depth 9, 15, 35, 44; grade beams 16–17, 20, 47; in high velocity wave action area 13, 15–17, 44–45; load bearing walls 9, 36; piers 9–10, 20, 36, 47; piles 9–10, 13, 15–16, 18–20, 36, 44, 46–47; posts 9–10, 16, 20, 36, 47
Freeboard 3, 38
Fuel storage tanks; design and construction guidelines for 25

Garages; design and construction guidelines for 27, 50–51
General scour 3
Grade beams; design guidelines for 16–17, 20, 47

H-section piles; materials requirements for 21–22

Heating systems; design and construction guidelines for 25
High risk flood hazard areas 37; alluvial fan areas 11, 38; defined 3; design and construction guidelines for 11–12; erosion prone areas 12, 39; flash flood areas 11, 38–39; high velocity flow areas 3, 12, 40; icejam and debris areas 4, 12, 40; mudslide areas 11–12, 39; siting in 8
High velocity flow areas; defined 3; design and construction guidelines for 12, 40
High velocity wave action areas; defined 3–4; design and construction guidelines for 12, 40–45; elevation requirements for 8, 13, 43; foundation requirements 13, 15–17, 44–45; identification of 13
Highest adjacent grade 4
Historic structure 4
Human intervention 4
HVAC; design and construction guidelines for 25, 49
Hydraulic elevator 49–50
Hydrodynamic loads 4
Hydrostatic loads 4

Icejam 4
Icejam and debris areas; design and construction guidelines for 12, 40
Impact loads 5

Knee braces 17

Landslide 39
Levees 4, 48
Load bearing walls 9, 36
Load combination 6
Loads 6, 10, 31–32; combinations 6; debris impact loads 1–2; hydrodynamic loads 4; hydrostatic loads 4; impact loads 5; piles 18–19; wave loads 6, 42
Local scour 4
Lowest floor 4

Mangrove stands 4, 43
Manufactured housing 29
Masonry; materials requirements for 22
Masonry walls; design requirements 18
Mat foundations; in high velocity wave action areas 13
Materials 3, 21–22, 47–48
Mats; design guidelines for 20
Metal connectors and fasteners; materials requirements for 21, 47

INDEX

Mitigation directorate 29
Mud flood 4, 30, 39
Mudflow 4, 30, 39
Mudslide 4, 30, 39
Mudslide areas; design and construction guidelines for 11–12, 39

National Flood Insurance Program (NFIP) 29, 30, 40
NAVD 4
New construction 4
New structure 4
NGVD 4
Non-engineered openings; in enclosures below design flood elevation 10, 36–37
Non-erodible soil 4
Non-load bearing walls 10, 31–32
Non-structural fill 9

Obstruction 4
One-hundred year flood 4, 31

Patios; design guidelines for 17, 26–27, 50
Piers; design guidelines for 9–10, 20, 36, 47
Pile capacity 46
Pile caps 19, 44
Piles; defined 4; design guidelines for 9–10, 18–20, 36, 46–47; in high velocity wave action areas 13, 15–16, 44; materials requirements for 21–22; minimum penetration 19; pile capacity 46; pile caps 19, 44; spacing 19; splicing 20, 47; supporting soil 19, 46–47
Plumbing; design and construction guidelines for 25, 49
Porches; design guidelines for 26–27, 50
Posts; design guidelines for 9–10, 16, 20, 36, 47
Precast concrete piles 18
Prestressed concrete piles 16
Prolonged contact with floodwaters 4
Protective works 32; in alluvial fan areas 11; in erosion prone areas 12; in flash flood areas 11, 39; in high velocity flow areas 12; in icejam and debris areas 12; in mudslide areas 12

Rafts; design guidelines for 20
Ramps; design and construction guidelines for 26, 50
Rapid drawdown 5
Rapid rise 5
Regulatory floodway 5
Reinforced concrete columns 16, 20, 44–45
Reinforced masonry columns 16, 20
Relocation 5
Revetments 17

Sand dunes 5, 43
Sanitary systems; design and construction guidelines for 25
Scour 5
Seawalls; defined 5; design requirements for 17
Shear walls; defined 5; in high velocity wave action areas 13, 15
Shield 5
Siting 8; in flood hazard areas 8, 13, 33; in floodways 8, 33–34; in high risk flood hazard areas 8; in high velocity wave action area 13, 43
Slabs; design guidelines for 20
Soil; pile design 19, 46–47
Special flood hazard areas 5, 30–31, 48
Spread footings; in high velocity wave action areas 13
Stairs; design and construction guidelines for 26, 50
Start of construction 5
Steel H piles 16
Steel HP-section piles 18
Steel piles 16, 18, 19
Steel rods; for cross-bracing 17
Stillwater depth 5, 14
Stillwater elevation 5
Storage tank 5
Structural fill 5, 9, 35–36
Structural steel; materials requirements for 21–22, 47
Structure framing; in high velocity wave areas 16
Structures; accessory structures 26–27, 50–51; chimneys 27, 51; classification of 7, 32–33; decks 17, 26–27, 50; defined 5; electrical elements 24–25, 49; elevators 26, 49–50; fireplaces 27, 51; flood control structure 3; floodproofing 2, 3, 5, 22–23, 48–49; garages 27, 50–51; identifying floodprone structures 6–7, 32; mechanical, heating, ventilation, and air conditioning 25, 49; patios 17, 26–27, 50; plumbing 25, 49; porches 26–27, 50; stairs and ramps 26, 50; utilities 23–24, 49–50
Substantial damage 5
Substantial improvement 5
Substantial repair 6
Substantially impermeable 5
Sump pumps 22

Timber; materials requirements for 22, 48
Timber piles 16, 18, 19, 44
Traction elevator 49–50

Utilities; design and construction guidelines for 23–24, 49–50

57

V zone 5, 31
Ventilation systems; design and construction guidelines for 25, 49
Vertical structural systems; design requirements for 18–20, 46–47

Walls; breakaway 1, 9, 17, 36, 45–46; concrete 18; floodwall 3; load bearing 9, 36; masonry 18; non-load bearing 10, 31–32; seawalls 5, 17; shear walls 5, 13, 15
Water surface elevation 5
Watershed 5

Wave crest elevation 5, 31
Wave height 5, 13, 14
Wave loads 5, 42
Wave runup 5, 14
Wave runup elevation 5, 13, 31
Waves 5; damage from 41–43; high velocity wave action areas 8, 12, 13, 40–45; wind-generated 41
Wet floodproofing 5, 23, 49
Wood; materials requirements for 22, 48
Wood piles; in high velocity wave areas 16, 44
Wood posts 16